Weather Hazard Warning Application in Car-to-X Communication

Attila Jaeger

Weather Hazard Warning Application in Car-to-X Communication

Concepts, Implementations, and Evaluations

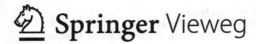

Attila Jaeger
Darmstadt, Germany

Darmstädter Dissertation
Dissertation at Technische Universität Darmstadt, 2015

D17

OnlinePlus Material to this book can be available on
http://www.springervieweg.de/978-3-658-15316-8

ISBN 978-3-658-15315-1   ISBN 978-3-658-15316-8   (eBook)
DOI 10.1007/978-3-658-15316-8

Library of Congress Control Number: 2016948251

Springer Vieweg
© Springer Fachmedien Wiesbaden 2016

Printed on acid-free paper

This Springer Vieweg imprint is published by Springer Nature
The registered company is Springer Fachmedien Wiesbaden GmbH
The registered company address is: Abraham-Lincoln-Strasse 46, 65189 Wiesbaden, Germany

Acknowledgment

First of all, I would like to thank my supervisor Prof. Dr.-Ing. Sorin A. Huss for his open minded and ongoing advice in technical and academic questions. His guidance and support has made this thesis possible. Secondly, I am grateful to Prof. Dr.-Ing. Ralf Steinmetz for being the second assessor.

I thank my colleagues at the Integrated Circuits and Systems Lab, Alexander Biedermann, Gregor Molter, Thomas Feller, Marc Stöttinger, Zheng Lu, Tolga Arul, Qizhi Tian, and Carsten Büttner. Special thanks are dedicated to Maria Tiedemann, who always provides a helping hand as well as kind and experienced support not limited to administrative matters. Additionally, I thank Tom Aßmuth, who introduced me to the most amazing sport, ever. It was a pleasure to cooperate, discuss, and share the daily work life with all of you.

I gratefully acknowledge the support of the Adam Opel AG, which provides the possibility to participate large scale Car-to-X field operational research projects to me. A lot of thanks to the respective colleagues at the Advanced Engineering Department, Bernd Büchs, Harald Berninger, Steffen Knapp, Jochen Stellwagen, and Marco Petrick for guiding and supporting my work within the projects. Thanks to the numberless further colleagues within sim^{TD} and DRIVE C2X for leading these projects to such a success.

Special thanks to Hagen Stübing for our time and discussions on technical and academic topics at the University, at Opel, and within sim^{TD}. Moreover, I am thankful for reviewing this thesis.

Thanks to Dennis, Nora, and David for your support and review of this work. Additional thanks to my girlfriend Sara for encouraging and supporting me all the time.

Last of all, I would like to give thanks to my brother and sisters, Camillo, Marie-Belle, Vivianne, and Dihia as well as, in loving memory, to my mother, Brigitte.

Danke euch allen!

Attila Jaeger

Kurzfassung

Car-to-X Kommunikation ist eine der vielversprechendsten Technologien um den Verkehrsfluss zu optimieren und die Verkehrssicherheit zu erhöhen. Dabei tauschen Fahrzeuge über eines dynamischen WLAN Netzwerk untereinander und mit Infrastrukturstationen Nachrichten aus. Diese Nachrichten können Informationen über die aktuelle Mobilität eines Fahrzeugs, also dessen Position, Geschwindigkeit und Fahrtrichtung, enthalten oder über eine spezifische Gefahrenstelle, wie bspw. einen Gegenstand auf der Fahrbahn, ein stark bremsendes Fahrzeug oder ein Stauende informieren. Basierend auf diesen Daten sind Anwendungen möglich, die durch adaptive Navigation oder Steuerung der Ampelphasen kooperativ den Verkehrsfluss verbessern oder den Fahrer über Gefahren informieren und so einen aktiven Beitrag zur Verbesserung der Verkehrssicherheit leisten.

In dieser Arbeit wird ein umfassendes Konzept für eine Anwendung, die über lokale Wettergefahren, wie bspw. dichten Nebel, Seitenwinde oder Glatteis, informiert, entwickelt und vorgestellt. Dabei werden, auf der einen Seite, im Fahrzeug verfügbare Sensordaten ausgewertet um Rückschlüsse auf die Wettersituation zu ermöglichen. Die so erlangten Informationen werden mittels Car-to-X Kommunikation mit anderen Fahrzeugen und Infrastrukturstationen geteilt. Andererseits werden Informationen, z. B. auf Grundlage von Wettermessstationen, von der Infrastruktur Verkehrsteilnehmern zur Verfügung gestellt. Dadurch kann, basierend auf einer breiten Datengrundlage, der Fahrer zuverlässig und gezielte über gefährliche Wettersituationen informiert und so kritische Fahrsituationen vermieden werden.

Offene Kommunikationssysteme, wie die Car-to-X Kommunikation, machen zwar vielversprechende Anwendungen möglich, sind aber auch angreifbar. In dieser Arbeit werden deshalb außerdem neue Konzepte zur Erkennung von Angriffen und Absicherung der Kommunikation entwickelt und vorgestellt. Mit diesen Gegenmaßnahmen wird eine Störung oder ein Missbrauch des Systems erschwert, so dass die Zuverlässigkeit und das Vertrauen in die Technologie sichergestellt werden.

Um die Umsetzbarkeit der vorgestellten Ansätze zu demonstrieren, wurde die entwickelte Anwendung und ausgewählte Konzepte im Rahmen der groß angelegten

Car-2-X Feldtests sim^{TD} und DRIVE C2X realisiert. So kann, basierend auf den Daten der umfassenden Versuche, die Effektivität der präsentierten Ansätze und der positive Einfluss der Anwendung auf die Verkehrssicherheit gezeigt werden.

Abstract

Car-to-X communication is considered as the key technology to optimize traffic efficiency and towards a significant increase of road safety. Therefore, based on a dynamic wireless network, vehicles exchange messages among each other and with stationary infrastructure stations. These messages, on one hand side, contain information about the current mobility of the vehicle, i. e., its position, speed, and direction of driving, and, on the other hand side, notify about a specific dangerous situation, such as an obstacle on the road, a hard braking vehicle, or the end of a traffic jam. Based on these data, applications are facilitated that cooperatively enhance traffic flow by, e. g., adaptive navigation or adjusted traffic light phases, or that actively improve traffic safety by informing the driver about local dangers.

Within this thesis, a comprehensive concept for an application notifying about local weather caused dangers, such as dense fog, strong cross winds, or slippery roads, is developed and presented. Therefore, on one hand side, data provided by in-vehicle sensors are evaluated in order to conclude on the current weather situation. This obtained knowledge is shared with other vehicles and stationary infrastructure stations using Car-to-X communication. On the other hand side, infrastructure stations provide information gained from, e. g., weather measurement stations, to road participants. In doing so, based on a comprehensive data basis, reliable driver notifications about specific dangerous weather situations are facilitated, such that critical driving situations might be avoided.

Open communication systems just like Car-to-X communication may enable promising applications but are vulnerable towards several attacks. Therefore, within this thesis, additional novel concepts to detect attacks and protect the communication are developed and presented. By applying these countermeasures disturbance or targeted misuse of the system is impeded, such that a reliable and trustworthy technology is facilitated.

In order to prove the usability of the presented approaches, the developed application and selected concepts are implemented and deployed within the context of the large scale Car-to-X field operational trials sim^{TD} and DRIVE C2X. Hence, based on the data gained within extensive tests, effectiveness of the presented approaches and the positive impact of the application on road safety are demonstrated.

Contents

List of Figures

List of Tables

Acronyms

ABS	Anti-lock Braking System
ACC	Adaptive Cruise Control
ADAS	Advanced Driver Assistant Systems
AES	Advanced Encryption Standard
API	Application Programming Interface
AR	Augmented Reality
ASIL	Automotive Safety Integrity Level
ASN.1	Abstract Syntax Notation One
AU	Application Unit
C2C	Car-to-Car (communication)
C2C-CC	Car 2 Car Communication Consortium
C2I	Car-to-Infrastructure (communication)
C2X	Car-to-X (communication)
CA	Certificate Authorities
CAM	Cooperative Awareness Message
CAN	Controller Area Network
CCU	Communication and Control Unit
CDD	Common Data Dictionary
CIS	Central ITS Station
CSMA/CA	Carrier Sense Multiple Access with Collision Avoidance
DCC	Decentralized Congestion Control
DENM	Decentralized Environmental Notification Message
DGPS	Differential Global Positioning System
DLE	Delta-Lambda-Equator
DoS	Denial-of-Service
DSRC	Dedicated Short Range Communications
DWD	Deutscher Wetterdienst
EC	Elliptic Curve
ECC	Elliptic Curve Cryptography
ECDSA	Elliptic Curve Digital Signature Algorithm

ECIES	Elliptic Curve Integrated Encryption Scheme
EGNOS	European Geostationary Navigation Overlay System
ESP	Electronic Stability Program
ETSI	European Telecommunications Standards Institute
Euro NCAP	European New Car Assessment Programme
FPGA	Field Programmable Gate Array
GIDAS	German In-Depth Accident Study
GNSS	Global Navigation Satellite System
GPS	Global Positioning System
HMAC	Hashed Message Authentication Code
HMI	Human Machine Interface
HMM	Hidden Markov Model
HTC	Hessian Traffic Center
HUD	Head-up Display
I2C	Infrastructure-to-Car (communication)
IGLZ	Integrierte Gesamtleit Zentrale of the City of Frankfurt am Main
ITS	Intelligent Transport Systems
IVI	In-Vehicle Information Message
JNI	Java Native Interface
KAF	Keep-Alive Forwarding
KDF	Key Derivation Function
LAN	Local Area Network
LDM	Local Dynamic Map
lidar	light detection and ranging
LLCF	Low Level CAN Framework
LoS	Line-of-Sight
MAC	Message Authentication Code
NIST	National Institute of Standards and Technology
NTP	Network Time Protocol
OBU	On-Board Unit
OSGi	Open Services Gateway Initiative
PDU	Protocol Data Unit
PKI	Public Key Infrastructure
PoI	Point-of-Interest
PoTi	Position and Time
PVDM	Probe Vehicle Data Message
QoS	Quality-of-Service
radar	radio detection and ranging

RDS-TMC	Radio Data System – Traffic Message Channel
RIS	Roadside ITS Station
RSA	Rivest, Shamir, and Adleman
RSSI	Received Signal Strength Indicator
SAM	Service Announcement Message
SHA	Secure Hash Algorithm
SPAT	Signal Phase and Timing Message
SVDF	Secret Value Derivation Primitive
SWIS	Straßenwetter-Informationssystem
TCS	Traction Control System
TMC	Test Management Center
TOPO	Road Topology Message
TTC	Text-to-Speech
UMTS	Universal Mobile Telecommunications System
UTM	Universal Transverse Mercator coordinate system
VANET	Vehicular Ad-Hoc Network
VAPI	Vehicle API
VDP	Vehicle Data Provider
VEBAS	Vehicle Behaviour Analysis and Evaluation Scheme
VII	Vehicle Infrastructure Integration
VIS	Vehicle ITS Station
WGS 84	World Geodetic System 1984

1. Introduction

Car-to-X communication is considered as the next major step towards a significant increase of road safety and traffic efficiency. For this purpose, vehicles connect to each other by building up a decentralized wireless ad-hoc network in order to jointly facilitate a various set of advanced and novel cooperative applications.

The term *Car-to-X* (C2X) thereby, summarizes *Car-to-Car* (C2C) communication, i. e., the communication between vehicles, as well as *Car-to-Infrastructure* (C2I) communication, i. e., the communication from vehicles to any kind of infrastructure components, and *Infrastructure-to-Car* (I2C) communication, i. e., the communication from any kind of infrastructure components to vehicles, respectively.[1]

One of the most promising applications is the *Weather Hazard Warning*, concerning weather related events, like dense fog, heavy rain or snowfall, aquaplaning, strong cross winds, or ice on the road. Within the context of this thesis, a novel and comprehensive concept for weather event detection, message protection and validation, information maintenance, and driver notification is developed, implemented, and evaluated.

1.1. Motivation

By enforcing a strict Road Safety Policy set by the European Commission [ECRSP], road deaths in the European Union has dropped from 54 000 to 30 500 by nearly 50 % in the years 2001 to 2011 [ECSAD]. To achieve the next ambitious goal to lower road deaths until 2020 down to about 15 500, new technologies are needed. Cooperative *Intelligent Transport Systems* (ITS) based on Car-to-X communication are considered as a key technology to achieve a next major breakthrough towards a significant improvement of active safety and traffic efficiency.

---

[1] Sometimes, especially in non-European literature, the noun *Car* is replaced by *Vehicle*, which leads to *Vehicle-to-X* (V2X), *Vehicle-to-Vehicle* (V2V), *Vehicle-to-Infrastructure* (V2I), and *Infrastructure-to-Vehicle* (I2V), respectively. However, in the context of this thesis the designation *Car* is used uniformly.

Based on data provided by the *German In-Depth Accident Study* (GIDAS), circumstances concerned by C2X use-cases and leading to traffic accidents with injured humans are investigated within context of the Car-to-X research project sim^{TD} [SIMTD09m]. Accordingly, about 30 % of road accidents occur in correlation with adverse weather conditions. Moreover, within about 4 % of the cases such weather conditions are determined as the only source leading to the accident. Hence, beside violated traffic signs, hard braking, and crossing traffic or left turning, bad weather conditions are identified as one of the major reasons for road accidents.

Nevertheless, regarding weather conditions such as, e. g., precipitation, road adhesion condition, reduced visibility, and strong wind, no sufficient solution is currently available, even if these conditions are affecting road safety sustainable [ELL83] [SWOV12]. These use-cases might be regarded within a Car-to-X application, the *Weather Hazard Warning*.

Within the Weather Hazard Warning application, the driver will get notified about upcoming local weather conditions. This way, a timely informed driver might react more appropriate to the changing road conditions. Thus, the risk and probability of accidents might be reduced.

In fact, to facilitate trustworthy weather notifications, an almost entire and reliable knowledge about the current local weather condition is obligatory. Therefore, the Weather Hazard Warning application has to regard as many weather relevant information as possible and, preferable, from multiple independent different sources.

Currently, a large approved infrastructure of sophisticated and well equipped weather measurement stations is already established. These stations are installed, maintained, and consulted by, e. g., local, national or international meteorological services. Additionally, present vehicles already carry a multitude of sensors, like temperature or rain fall sensors, suitable to identify the local weather conditions.

So far, however, observations from the infrastructure of measurement stations are only available to associated services and measurements by vehicle local sensors are provides to the local vehicle, only. Consequently, the Weather Hazard Warning application has to bridge three information gaps by means of C2X communication:

1. Knowledge about road safety relevant weather conditions provided by measurement stations has to be made available to vehicles on the road.

2. Information about weather hazards available in vehicles has to be made available to adjacent vehicles in the near environment.

3. Weather relevant data gathered by local sensors in the large number of vehicles spread all over the roads, has to be regarded by meteorological

services to extend knowledge based on established weather measurement stations.

Accordingly, within the context of this thesis, an application based on Car-to-X communication is developed to overcome the mentioned gaps and providing approaches, concepts, and reference implementations for ongoing standardization and the upcoming market launch of C2X systems.

1.2. Weather Hazard Warning Application Overview

Car-to-X communication enables a various set of road safety and traffic efficiency related use-cases. Multiple C2X use-cases are summarized and categorized by the *European Telecommunications Standards Institute* (ETSI) in [ETSI09], whereas basic requirements are stated in [ETSI10b]. These include the mentioned major threats to road safety, i.e., traffic sign violation, hard braking, crossing traffic or left turning, and bad weather conditions.

However, today's vehicles are already equipped with traffic sign assistance based on navigation maps and cameras. Thus, countermeasures against traffic sign violation are currently established and might be widely spread in the near future. For hard braking situations, it has to be mentioned that time slots for driver reaction are very hard restricted. Such that, autonomous emergency brake assistance systems based on, e.g., camera, radar, or lidar, are already available since a few years, and will have a significant effect on road safety. In order to facilitate reliable cross traffic and left turn related use-cases based on C2X communication, almost every vehicle has to be equipped with the novel technology. Even by enforcing introduction of C2X system by governments and vehicle safety institutes, like the *European New Car Assessment Programme* (Euro NCAP), a sufficient penetration rate will not be achieved within foreseeable future.

Hence, within the large set of defined use-cases, the *decentralized floating car data*-based use-cases[2]

- precipitation
- road adhesion condition
- visibility condition
- wind problem

are most promising to enhance road safety significantly by utilizing C2X communication. These four use-cases concerning weather conditions affecting road safety

---

[2] According to [ETSI09] and [ETSI10b] these use-cases are referred to as UC014, UC015, UC016, and UC017, respectively.

sustainable shall be regarded together within one Car-to-X application. Thus, the driver gets notified about upcoming local weather events, which may sustainable impair the road safety. Within the context of this thesis, this application is referred to as the *Weather Hazard Warning* application.

This application obviously has to be regarded as a *Cooperative Road Safety* application and is categorized by current standardization accordingly [ETSI09] [ETSI10b].[3] Additionally, the use-cases of the *Weather Hazard Warning* may be consulted as source of information for supplementary route guidance related use-cases, e. g., *Traffic Information and Recommended Itinerary*, *Enhanced Route Guidance and Navigation*, or even *Limited Access Warning and Detour Notification*.[4] The supplementary information about road weather conditions may have a sustainable effect to the results of these use-cases leading to advanced route guidance. Thus, the Weather Hazard Warning might be also regarded as a *Cooperative Traffic Efficiency* related application.

The focused goal for the Weather Hazard Warning application is to notify the driver timely about possibly dangerous road conditions ahead before the vehicle will reach the situation. In doing so, the driver might adapt his driving behavior in time, will be prepared, and, consequently, might react appropriate to the changing road conditions leading to a reduced number of driving failures.

Therefore, the Weather Hazard Warning application involves multiple sources for weather relevant information. Thus, a comprehensive and reliable view on the road weather conditions is achieved.

On the one hand, there is an already established and well-proven weather measurement station infrastructure maintained by, e. g., meteorological services like the *Deutscher Wetterdienst* (DWD)[5]. Such a station consists of a multitude of highly sophisticated sensors already observing weather conditions, like temperature, wind speed and direction, rainfall or snowfall, range of vision, etc. These weather relevant data might be made available to traffic participants via C2X communication technology.

On the other hand, today's vehicles are equipped with a composition of suitable sensors such as temperature, light, or rain sensors. In addition, driver action such as activated front shield wipers or rear fog lights may give evidence on current

---

[3] In contrast to [ETSI09], within [ETSI10b] this category is referred to as *Active Road Safety*. However, since the document excludes any *active* behavior and, moreover, the respective use-cases have to be considered as highly *cooperative*, the designation *Cooperative Road Safety* is adopted in context of this thesis.

[4] Use-cases UC020, UC021, and UC022, respectively.

[5] The German federal meteorological institute [DWD].

local weather conditions. Exchanging these data between vehicles significantly enhances an applications detection ability and reliability compared to applications considering own vehicle's local sensors, only.

Moreover, vehicles on the road may be regarded as a network of mobile sensor nodes, providing valuable information about the current weather or climate in a specific region. Hence, gathered local vehicle sensor data are of high interest for meteorological institutes. Combining weather information provided by vehicles together with conventional sources like, e. g., satellite observations or weather measurement stations, supports meteorological institutions to create a more accurate weather view. Thus, their forecasts are significantly enhanced [SEI08]. Consequently, within different C2X research projects weather related public agencies, e. g., the meteorological service DWD, the German federal road agency *Bundesanstalt für Straßenwesen* (BASt), or local traffic centers like, e. g., *Hessen.mobil*, are involved. But also institutes from the private sector, e. g., the *Meteomedia AG* [METEO] showed great interest in cooperating with a C2X system.

According to the great advantages for improving road safety and traffic efficiency within the last years different approaches for a *Weather Hazard Warning* application are extensively investigated in C2X related and comparable research projects.

An early approach to deliver road weather related notifications to vehicles is pronounced in [SMI02]. This rudimentary system is designed to deliver individual calculated forecast to road participants. Due to the individuality of these forecasts, they are only available to registered subscribers which have to be equipped with special devices applicable for this system only. Additionally, due to its centralized structure, a dedicated center has to calculate the forecast depending on continuously tracked detailed and personalized information about current vehicle position and trip destination. Consequently, this system is hard to establish in large scale and, moreover, implies a huge challenge to driver's privacy.

Another strategy to deliver weather notifications into vehicles is based on the *Radio Data System – Traffic Message Channel* (RDS-TMC) [IEC00]. Thereby, information is embedded into conventional radio broadcasts and, thus, transmitted to all receivers within communication range of the respective radio stations. This includes a huge amount of vehicles, even those who are far away and not affected by the danger. However, besides the frequently complained issues with outdated notifications and traffic information within RDS-TMC, it is shown that the system is purely secured against message injection and forgery [BB07] [SYM10]. Furthermore, within this approach there is no opportunity to expand data basis of meteorological services by regarding vehicles as mobile sensors.

In contrast, the North American initiative *Vehicle Infrastructure Integration* (VII)[6] starts to advocate *Dedicated Short Range Communications* (DSRC), an early precursor to the current Car-to-X technology, in order to distribute weather related information [PM07]. Thereby, VII focuses towards enhanced guidance of road maintenance vehicles [PMC+08], like snowplows, and only incidentally towards in-vehicle warnings and information [SPTG09]. However, within VII weather related information originated by conventional sources, e. g., measurement stations, weather radars, or satellites, are combined together with probe data collected by vehicles to achieve an advanced view on weather conditions.

Based upon afore made experiences, a novel and comprehensive concept for the Weather Hazard Warning application based upon C2X technology is developed by the author of this thesis. An overview of the concept is announced in [HJS11], whereas selected parts are shortly outlined in [SJ10], and, especially its integration into a large scale field operational test, is described in [JH11]. Thereupon, approaches, components, techniques, security and privacy considerations, and driver notification strategies of the Weather Hazard Warning application are detailed and discussed within context of this thesis. Moreover, reference implementations and evaluation results achieved in context of current research project, i. e., sim^{TD} and DRIVE C2X, are presented.

Within the developed application local sensor data are captured and processed in order to recognize local weather situations. Consequently, each vehicle passing a weather event detects weather conditions independently and distributes gathered information to other vehicles. In doing so, all road participant work together cooperatively in order to enhance the overall knowledge. Thereby, the C2X communication system provides messages formats, communication technology, and advanced message distribution algorithms to the application.

Based upon received information, the application is facilitated to notify the driver about upcoming events. Therefore, data gathered from own vehicle sensors, originated from other vehicles, and provided by infrastructure-based services are combined. Thus, the presented application addresses all three major information gaps as previously stated in Section 1.1.

---

[6] After renaming *Vehicle Infrastructure Integration* to *IntelliDrive*^{SM} in 2009, as of 2011 the initiative is referred to as *Connected Vehicle Research* by the *Research and Innovative Technology Administration* of the U.S. Department of Transportation [RITA].

1.3. Chapter Outline

Following this introduction, Chapter 2 provides a short introduction into Car-to-X communication, possible use-cases, and related standards. Subsequently, an overview of the overall system architecture, different station types and their components is briefly described. Since the Weather Hazard Warning is based on a cooperative approach, detected situations have to be communicated to other road participants. Accordingly, selected C2X messages formats, their composition, and their intended use in context of this thesis is detailed. Additionally, several applied distribution algorithms are outlined. Finally, several approaches to represent geographical coordinates and areas are presented and discussed.

The structure of the following next three chapters roughly follows the path of passing weather relevant data through the C2X communication system.

Hence, Chapter 3 starts with a description on how the environmental weather situation is detected by the application. Based upon the fundamentals detailed in Chapter 2, the vehicle-based detection is described by, firstly, outlining the processing of available vehicle sensor and state data. Thereon, two approaches to detect potential dangerous adverse weather situations and events within vehicles are described in detail and discussed. In addition, an overview on processing data and detection of weather situations on infrastructure side, exploiting weather measurement stations, is outlined at the end of the chapter.

After a message has been created by a detecting station, it will be received by appropriate vehicles. Since Car-to-X communication is inherit vulnerable to several attacks against message's content, Chapter 4 starts with countermeasures by means of plausibility verification of message's content[7]. Hence, only reliable messages might be provided to the application. Subsequently, it is detailed how received information is processed by the Weather Hazard Warning application. Therefore, the aggregation of weather situations regarding their area and detection reliability is described and suitable techniques and algorithms are stated. Finally, an outline regarding the management of known situations and events is provided.

In Chapter 5, the final step of a current C2X application, i. e., how a notification of the driver is realized, is detailed. Thereby, methods to identify relevant weather events are discussed. Additionally, presenting strategies by means of consulted instruments, notification intensity, and warning levels are outlined.

---

[7] An introduction into possible attacks on Car-to-X communication and further countermeasures against falsified messages and there implementation and evaluation are detailed in Appendix A.

At this point, the complete path of passing weather relevant information through the C2X communication system is outlined. Consequently, Chapter 6 details reference implementations regarding afore proposed methods, components, and the Weather Hazard Warning application. Thereby adaptions, variations, and integration details for deployment within a system architecture for field operational tests are stated.

Subsequently, in Chapter 7 evaluation results based on presented implementations of the presented methods and components gained during real world test trials are detailed. Moreover, especially for the Weather Hazard Warning applications deployed in large scale field operational test, in-depth evaluations on the reliability of the situation detection and the driver acceptance are presented.

Finally, Chapter 8 summarizes the thesis and concludes with a brief outlook.

2. Car-to-X Communication

Modern vehicles carry a large number of highly sophisticated sensors to provide a various set of comfort and active safety technologies. Hence, not restricted to upper class vehicles, there are, e. g., radar or lidar sensors to enable *Active Cruse Control* or *Active Braking*, one or even multiple cameras for *Road Sign Assistance*, *Pedestrian Recognition*, or *Lane Keeping Assistance*, as well as sonar sensors for *Blind Spot Warning* or *Parking Assistance*.

All these sensors provide measured data to applications running on the vehicle. Consequently, combining all available information, a precise assumption of the local environment around the vehicle is gathered, but limited to the respective range of the respective sensor. Sonar sensors detect objects in the direct surrounding of the vehicle, i. e., in a range of up to 4 meters. Typical automotive cameras available today, provide reliable detection up to a total distance of 80 meters. Finally, credible object detection with radar or lidar systems is achievable for a maximum distance of about 200 meters, if objects are in direct line of sight.

The major idea behind Car-to-X communication is to share local vehicle sensor data with other, potentially far away, vehicles by using wireless communication. Hence, by using multi-hop schemes, whereas vehicles forward received messages, information may be transmitted over several kilometers. Thus, the view on the local environment can be expanded significantly. Additionally, since information may be forwarded by intermediate vehicles, there is no need of a direct line of sight to achieve information about an object of interest. In Figure 2.1 different measurement ranges of selected vehicle sensors are depicted in comparison and complemented by the possible additional range of a C2X communication system.

The wide range of different C2X applications and use-cases is grouped by the *European Telecommunications Standards Institute* (ETSI) into three major categories [ETSI09], i. e.,

- Cooperative Road Safety,
- Cooperative Traffic Efficiency, and

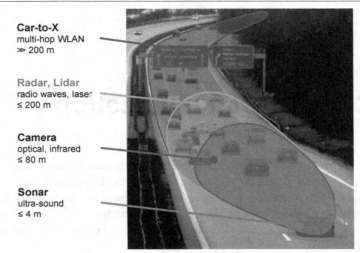

cf. [SJ10]

FIGURE 2.1.: Comparison of selected vehicle sensors with respect to utilized technologies and their respective maximum working range.

- Cooperative Local Services and Global Internet Services.

In recent years, within multiple research projects with field operational trials, e. g.,
DIAMANT, aktiv [AKTIV], sim^{TD} [SIMTD], or DRIVE C2X [DRIVE], almost the
entire spectrum of applications was specified, implemented, and examined. Based
on gained experiences, ETSI consolidated C2X applications and defined functional
requirements for a *Basic Set of Applications* [ETSI10b].

Beside the weather related use-cases which are within the focus of this thesis, these
use-cases include, e. g.,

- the *Intersection Collision Warning* to avoid collisions with crossing traffic,
- the *Emergency Vehicle Warning* to notify about operating rescue or police
  vehicles, such that the driver may make room for them,
- the *Emergency Electronic Brake Lights* emphasizing hard braking vehicles in
  the front,
- the *Traffic Light Optimal Speed Advisory* suggesting optimal speed to pass
  traffic light series at a *green-wave*,
- the *In-vehicle Signage*, informing about current traffic rules, e. g., speed
  limitations, and notifying once they are violated, or
- the *Traffic Condition Warning* notifying about, e. g., an upcoming traffic
  congestion.

Additional use-cases, like a notification about crossing trains [JF14], are under investigation and development.

Within the following sections, a short overview on the C2X communication architecture, communication technology, selected C2X messages formats, as well as area and coordinate representation facilitating such advanced applications is provided.

2.1. Overall Car-to-X System Architecture Overview

Within Car-to-X communication vehicles exchange information between each other. In addition, stations placed along the roadside are exploited to enhance communication range and to connect infrastructure-based entities to the vehicles. Accordingly, a C2X communication system architecture consist of three different basic entity types [ETSI13a] as follows.

Vehicles are equipped with *Vehicle ITS Station* (VIS) to facilitate C2X communication. The VIS thereby consists out of a computation unit[8] hosting *applications* and *system components*. While applications are realizing C2X use-cases, system components are providing basic services required by other system components or applications, e. g., the vehicle's position and sensor data or message en- and decoding.

This computation unit is connected to an interface to the driver, e. g., a display with touch-screen, to enable interaction between the C2X system and the driver. This way, e. g., warning messages are displayed or driver input is gathered.

However, the unit additionally is connected to vehicle's internal data and local sensors. This is usually a connection to the vehicle's CAN bus. Thus, vehicle dynamics, local sensor data, vehicle state, and diagnostic data, e. g., vehicle speed, exterior lights state, wipers activation, occupied seats, etc., are available within the C2X system. Moreover, this includes a position provider, e. g., a GPS receiver, providing the vehicle's position and a reliable time synchronization source.

And finally, at least one C2X communication capable antenna, i. e.. at a frequency of 5.9 GHz, is connected to the computation unit. Hence, sending and receiving of C2X messages is facilitated.

Thereby, the communication range of individual VIS is enlarged by the mechanism of *multi-hop*. Hence, C2X messages are forwarded, i. e., received and retransmitted, by multiple stations until they reach their destination.

---

[8] In fact, within research projects usually multiple interconnected units are exploited for technical reasons. However, omitting technical details, these units might be regarded as one logical entity.

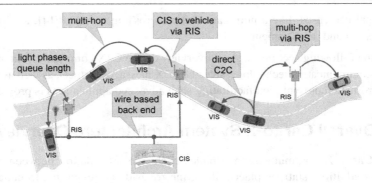

FIGURE 2.2.: The three different C2X station types, i. e., VIS, RIS, and CIS, and
selected communication scenarios between them.

In addition to vehicles, *Roadside ITS Stations* (RIS), which are placed along the
roadside, are involved in C2X communication. These stationary stations might
further enlarge the communication range, especially within poor frequented areas,
by repeating received messages.

Roadside stations might belong to a large network of stations, connected via a
wire-based back-end, or are attached to one or a small group of infrastructure
components, e. g., traffic lights. This way, messages might be forwarded wire-based
to adjacent RIS or a station might provide an interface to infrastructure components,
thus, e. g., traffic light phases are providing to vehicles or waiting queue length is
communicating to a traffic lights controller.

Obviously, these roadside stations do not require an interface to a driver or a
connection to vehicles data. In contrast, they are connected to the infrastructure
component or the interconnection network.

As a third participating type of entities, a various set of different infrastructure-
based service providers and agencies, like local traffic centers, may be connected
to the C2X system. Such *Central ITS Stations* (CIS) usually do not communicate
directly to road participants. Instead, they are connected to a RIS or a network of
multiple RIS, which again serve as an interface between infrastructure and vehicles.

These different station types[9] and selected communication links between them are
illustrated in Figure 2.2.

---

[9] Additional *Personal ITS Stations*, as, e. g., investigated within the research project AMULETT
[AMU], are intended, but are not further regarded in context of this thesis.

2.2. Frequency and Channel Allocation

Car-to-X communication is based on a wireless ad-hoc network between vehicles and other road traffic relevant participants. Such a wireless *Vehicular Ad-Hoc Network* (VANET) is exploited to exchange information between these participants and to enable a large set of road safety and traffic efficiency related applications.

To enable communication for such cooperative *Intelligent Transport Systems* (ITS), carrier frequencies in the range of 5.9 GHz are allocated on physical layer. Such a technology is specified as *IEEE 802.11p* in [IEEE10] and, based on this, standardized in [ETSI12] within the European Union. Accordingly, there are eight communication channels in four bands provided to ITS communication stations. Only three bands, i. e., ITS-5GA, ITS-5GB, and ITS-5GD, are intended for C2X communication systems as in the scope of this thesis. While ITS-5GD is reserved for future use, ITS-5GA is restricted to road safety related applications only and ITS-5GB is intended for traffic efficiency purposes, respectively.

The three regarded ITS-5G bands are divided into several channels of 10 MHz width. One physical channel is allocated to be the *control channel* (CCH) while the others are defined as *service channels* (SCH), i. e., channel SCH1 to SCH6.

In order to avoid interferences between individual channels and neighbored frequency bands, for each channel the per MHz power density is limited to -10 dBm, 13 dBm, or 23 dBm, i. e., 0.1 mW, 20 mW, or 200 mW, respectively.

The channel and G5-band frequency allocation, respective power density limits, and intended purpose are depicted in Figure 2.3

2.3. Car-to-X Messages

Car-to-X communication implies information exchange between different ITS stations, i. e., Vehicle ITS Stations, Roadside ITS Stations, and Central ITS Stations. Relevant information is wrapped in *C2X Messages* and transmitted by means of wireless communication.

There are three major message formats used in multiple C2X use-cases. Firstly, the periodically sent *Cooperative Awareness Message* (CAM) as specified in [ETSI11b]. Secondly, the event-driven *Decentralized Environmental Notification Message* (DENM) as specified in [ETSI13c]. And thirdly, the *Probe Vehicle Data Message* (PVDM) containing collected sensor measurements.

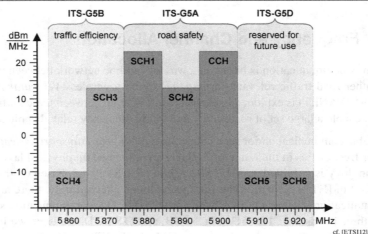

cf. [ETSI12]

FIGURE 2.3.: Allocated frequencies for ITS channels and G5-bands, their intended
purpose, and power density limits.

In addition, supplementary use-case specific messages are used, e. g., the *Signal
Phase and Timing Message* (SPAT) wrapping information about traffic light phases,
the *Road Topology Message* (TOPO) containing detailed information about the
local road structure, the *In-Vehicle Information Message* (IVI) informing about
local traffic signs, or the *Service Announcement Message* (SAM) sent by RISs to
notify about their provided services. However, these Messages are not involved in
the Weather Hazard Warning application and, hence, not further detailed.

In the following sections the three major message formats are detailed. Subsequently, a short overview on message distribution algorithms is stated.

2.3.1. Cooperative Awareness Message

The *Cooperative Awareness Message* (CAM) is the basic message in Car-to-X
communication. It facilitates multiple C2X use-cases, e. g., *Emergency Vehicle
Warning*, *Slow Vehicle Indication*, *Intersection Collision Warning*, or *Motorcycle
Approaching Indication*. CAMs contain basic information about the current ITS
station's position and state. Due to the mobility within C2X communication, CAMs
are updated periodically. This way, other road participant are notified about the
presence of the originating *Vehicle ITS Station* or *Roadside ITS Station*.[10]

These notifications are relevant to all ITS stations in the direct neighborhood.

---

[10] Obviously, the state of a roadside stations may basically consist out of the stations position, only.
Hence, the description of CAMs focuses on messages sent by vehicles.

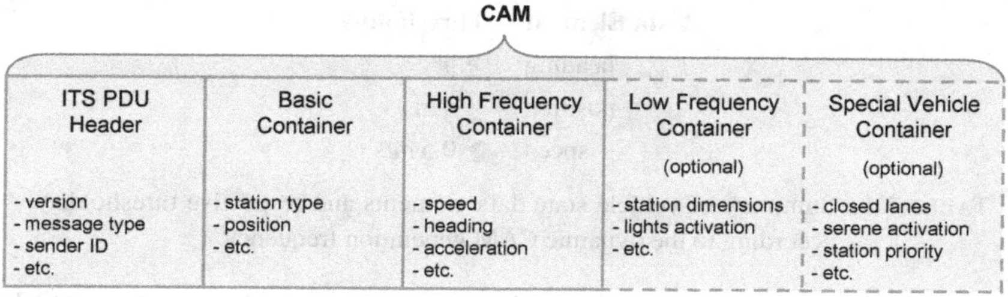

CAM

ITS PDU Header	Basic Container	High Frequency Container	Low Frequency Container (optional)	Special Vehicle Container (optional)
- version - message type - sender ID - etc.	- station type - position - etc.	- speed - heading - acceleration - etc.	- station dimensions - lights activation - etc.	- closed lanes - serene activation - station priority - etc.

cf. [ETSI11b]

FIGURE 2.4.: General structure of the *Cooperative Awareness Message*, assembled out of the common ITS header followed by the mandatory and the optional containers.

Thereby, *single-hop* communication is applied, while forwarding mechanisms are not regarded. Hence, CAMs are broadcasted to every station in direct communication range, only.

According to [ETSI11b], as every other ITS message, CAMs contain the common *ITS PDU Header*, providing basic information about the message type, the version, etc. The mandatory *Basic Container* indicates the type of the ITS station, i. e., VIS or RIS, and the geographical position.

In case of a vehicle station, additional containers are added. The mandatory *High Frequency Container* contains all rapidly changing vehicle state data, e. g., speed and heading. They are updated within every CAM. Slowly changing or static vehicle state data, e. g., exterior light activation, open doors, and vehicle dimensions, do not have to be updated frequently. Consequently, the optional *Low Frequency Container*, wrapping these data elements, is not transmitted in every CAM. Special vehicle types, e. g., public transport, road works, or operating emergency vehicles, may add a respective *Special Vehicle Container*, containing according additional vehicle specific state information.

In Figure 2.4 this general structure of a CAM is summarized. A detailed description of the *Cooperative Awareness Message* is given in [ETSI11b].

As mentioned, CAMs are updated and sent periodically. During field operational tests, different static frequencies, i. e., 1 Hz, 2 Hz, or 10 Hz are implemented and evaluated. However, since CAMs are intended to notify about the state of an ITS station, a new CAM must only be generated if the state has changed significantly. Hence, multiple criteria are specified whenever such a significant change is indicated [ETSI11b]. Accordingly, Table 2.1 summarizes vehicle state data elements and respective thresholds. These criteria are evaluated periodically and if one condition is met, a new CAM is triggered. Thereby, a minimum time interval t_{min} of

Data Element	Threshold
heading	$\geq 4°$
position	$\geq 4\,m$
speed	$\geq 0.5\,m/s$

TABLE 2.1.: Summary of vehicle state data elements and respective thresholds, according to the dynamic CAM generation frequency.

$$t_{min} = 100\,ms \tag{2.1}$$

between two consecutive messages, whereas CAM generation is omitted, is considered. Analogous, a maximum time interval t_{max} of

$$t_{max} = 1000\,ms \tag{2.2}$$

between two consecutive messages is considered. Thereupon, a new CAM is generated anyway.

Accessorily, the current channel load is regarded to encourage *Decentralized Congestion Control* (DCC). Hence, the minimum CAM transmission interval t_{min} may be increased if required by DCC. Such a dynamic CAM frequency facilitates as much CAM updates as needed to enable effective C2C use-cases and prevents flooding the channel with redundant information.

2.3.2. Decentralized Environmental Notification Message

The *Decentralized Environmental Notification Message* (DENM) is one of the most important messages in Car-to-X communication. It facilitates multiple C2X use-cases, e. g., *Road Works Warning*, *Emergency Electronic Brake Light*, *Traffic Jam Ahead Warning*, or *Weather Hazard Warning*. Thereby, a DENM contains detailed information about a specific dangerous situation[11]. Hence, road participants are informed timely about relevant hazardous events in the near environment. These explicit situation notifications may be generated from roadside stations located close to the situation, e. g., at a static or mobile road works place, from vehicles detecting the danger while driving nearby, or from central stations receiving data from external sources or other ITS stations.

A timely notification of affected vehicles requires communication over large distances. Since C2X communication technologies facilitate direct transmission over

---

[11] Within respective European standardization, no sharp distinction between the terms *event* and *situation* is made. In contrast, within context of this thesis, a single notification, as reported by a DENM, is denoted as *situation*, whereas the overall real world incident is denoted as *event*. Consequently, respective terms are adapted, accordingly.

Event	Cause Code	Sub-cause Type	Description
Dense Fog	018	001	visibility reduced – due to fog
Heavy Rain	019	001	precipitation – heavy rain
Heavy Snowfall	019	002	precipitation – heavy snowfall
Slippery Road	006	005	slippery road – ice on road
Aquaplaning	007	000	aquaplaning – *no sub-cause*
Strong Winds	017	001	extreme weather – strong winds

TABLE 2.2.: Summary of *Cause Codes* and *Sub-cause Type* exploited to encode weather events within DENMs generated by the Weather Hazard Warning application.

up to several hundred meters, only, *multi-hop* techniques are applied to distribute DENMs [C2C07]. Accordingly, messages are repeated by multiple receiving stations. Thereby, no specific destination address is provided. In contrast, *geocast* mechanisms are applied, whereas geographical destination areas are consulted, instead [MAI04]. This way, messages are forwarded towards the geographical destination area (*Store & Forward*) or, if already located inside the intended region, are kept available until the end of their validity duration (*Keep-alive Forwarding*).

According to [ETSI13c], as every other ITS message, DENMs contain the common *ITS PDU Header*, providing basic information about the message type, the version, etc. The mandatory *Management Container* contains essential situation information, e. g., an *ActionID* to refer to the situation, a *validity duration* specifying how long the message is repeated, a *transmission interval* defining how frequently the message is repeated, etc.

Detailed situation specific information is wrapped within the optional *Situation Container*. This includes, e. g., the situation type, the information quality, the situation position, etc.

Thereby, situation types are encoded according to *Transport Protocol Experts Group – Traffic Event Compact* (TPEG-TEC). Hence, [ETSI13b] specifies *Cause Codes* with *Sub-cause Type* based upon [ISO13][12]. A listing of Cause Codes with Sub-cause Type exploited by the Weather Hazard Warning application is summarized in Table 2.2.

The mentioned information quality denotes the reliability of the provided information. This reliability value is given as a certain level within the range of 0 to a

---

[12] Earlier versions of [ETSI13c] recommend cause codes according to [TPEG06]. This is updated during international and technologies overlapping harmonization efforts.

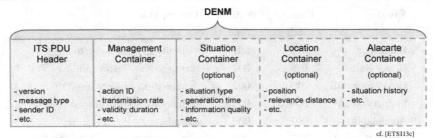

cf. [ETSI13c]

FIGURE 2.5.: General structure of the *Decentralized Environmental Notification Message*, assembled out of the common ITS header followed by the mandatory and the optional containers.

use-case specific maximum level. To determine the position of the situation, a geographical position in combination with a relevance distance is consulted, according to [ETSI13b][13].

The optional *Location Container* holds supplementing localization information, especially for moving situations, e. g., situation speed, situation position heading, etc. Finally, the *Alacarte Container* provides various optional additional use-case specific information, not included in other DENM containers.

In Figure 2.5 this general structure of a DENM is summarized. A detailed description of the *Decentralized Environmental Notification Message* is given in [ETSI13c].

For weather events it has to be considered that they are almost static and that a weather event affects an area regardless of which direction a vehicle approaches. Hence, no event mobility information or approaching traces are required in weather related DENMs. Based on the provided possibility to encode a geographical area, the Weather Hazard Warning expresses situations as a rectangle surrounding the hazardous area.

Since DENMs contain detailed information about local dangers, they facilitate data exchange between vehicles and from infrastructure-based services to vehicles. Thus, a DENM might be exploited to overcome the first and the second information gap as stated in Section 1.1.

---

[13] Only predefined relevance distances of, e. g., 100 m, 500 m, 1000 m, or 5000 m are selectable. This implies insufficient possibility to specify a geographical region. Earlier versions of [ETSI13c] provide the possibility to define a geographical event position, according to [ETSI10c]. Thereby, the area may be shaped as a circle, a rectangle, or an ellipse as further detailed in Section 2.5.1. Current standardization efforts tending to provide these advanced positioning themes, again.

2.3.3. Probe Vehicle Data Message

By providing essential data gathered by local vehicle sensors, the *Probe Vehicle Data Message* (PVDM) facilitates multiple C2X use-cases, e. g., *Road Works Information, Enhanced Route Guidance and Navigation, Traffic Condition Warning*, or *Weather Hazard Warning*. Thereby, use-case specific relevant vehicle sensor data, like vehicle speed and outside temperature, are gathered and stored locally in combination with positioning and timing information. Subsequently, accumulated batches of these data are packed in messages and sent from the vehicle via RISs to Central ITS Stations. These *profiles* extend the data basis of the respective infrastructure-based C2X use-case or application. Thus, vehicles are exploited as mobile sensors along the road, providing essential information. Especially the infrastructure-based parts of the Weather Hazard Warning application benefit from the availability of this additional knowledge. Thus, generation of more precise warning messages is facilitated.

Since PVDMs contain rarely processed data, the information need to be interpreted. Due to rather low processing power within vehicles, this extensive data evaluation is not performed on-board, but on the infrastructure side, only. Consequently, PVDMs are not intended to be transmitted to other vehicles, but are forwarded to the CIS, instead.

Suchlike messages are based upon [ISO09] and have been exploited within large scale field operational tests as, e. g., sim^{TD} [SIMTD09j]. However, standardization is still not finalized. Accordingly, a substantiate structure for PVDMs is proposed, by the author of this thesis, which is in compliance to the container-based specification of CAM and DENM.

As every other ITS message, PVDMs shall contain the common *ITS PDU Header*, providing basic information about the message type, the version, etc. Subsequently, a mandatory *Timing and Positioning Container* containing essential position and time information for each measuring point, like geographical position, generation time, and validity duration.

Accessorily, at least one optional use-case specific container is added. Data elements included in such a container depends on the respective C2X use-case. Hence, e. g., a *Traffic Data Container* wrapping, e. g., vehicle speed and heading, or a *Weather Data Container* wrapping, e. g., ambient temperature or rain intensity, may be consulted. In Figure 2.6 this proposed possible general structure of a PVDM is summarized.

As mentioned, probe vehicle data are intended to be sent to central stations, only.

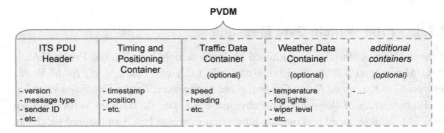

FIGURE 2.6.: Presumably general structure of the *Probe Vehicle Data Message*, assembled out of the common ITS header followed by the mandatory and the optional containers.

Thereby, roadside stations are exploited as gateways between the wireless C2X communication and the wire-based connected infrastructure side. Consequently, once a vehicle reaches the communication rage of a RIS, gathered measurements are packed into a PVDM and delivered to the forwarding roadside station.

However, a RIS may not be capable to forward every use-case specific container to an appropriate CIS. Hence, they may broadcast a *Service Announcement Message* (SAM) [ETSI14] containing detailed information about their facility to receive and forward data. Thus, passing vehicles are notified about which containers may be transmitted with the current PVDM towards the respective RIS.

Such a PVDM facilitates data exchange from vehicles to the infrastructure. Thus, by exploiting PVDMs the third information gap as stated in Section 1.1 might be addressed and reduced.

2.3.4. Message Distribution Algorithms

This section provides a short introduction to message distribution in C2X communication. Respective mechanism and algorithms are mentioned and briefly outlined. Detailed information may be achieved within relevant literature and standards.

Geocast Distribution Schemes

Within Car-to-X communication messages are transmitted wireless. In contrast to conventional networking schemes, in addition to sending a message to a specified receiver by its address, i. e., *unicast*, and sending to every receiver within communication range, i. e., *broadcast*, sending messages to a specified geographical area, i. e., *geocast*, is supported [C2C07], too. Within a geocast, a message may be forwarded by multiple *nodes* to a geographical region and delivered to receivers within that region [ETSI10d]. However, even for unicasts C2X specific conditions especially regarding the high mobility of the ad-hoc network have to be considered.

Hence, each node, i. e., a Roadside or a Vehicle ITS Station, maintains a *Location Table*, storing the position, identifier, and time information of each known node. Thus, if a message from an unknown node is received, a new entry is added to the Location Table. Otherwise, if a new message from a known node is received, respective position and time information are updated. In contrast, if no message from an individual node was received for a certain time period, the node's entry is removed from the Location Table.

On the other hand, each C2X message contains respective destination and transmitting information such as

- a target destination node's *identifier* and *position*,
- a target *destination area*, i. e., a geographical region,
- a *validity duration*, determining for how long the message shall be sent or forwarded, or
- a *hop-limit*, i. e., the maximum allowed number of intermediate forwarding nodes.

Based on the intended transmission scheme, not every of the enumerated information is required.

Accordingly, on reception of a message, the receiving ITS station evaluates the message's addressing and transmitting information and determines

1. if it is a destination for the message and
2. if the message shall be forwarded.

Thereby, the receiving node may be a destination due to its identifier or its location. On the other hand, if the node is a destination or if not, the message may be required to be forwarded. Thus, a node may be a destination and a forwarder at the same time. Thereby, it is considered, if the hop limit is reached or the validity duration is expired. Additionally, based on the location table, it is evaluated if there are possible receivers within communication range or if there is a route, from the current node to the destination. However, a survey of possible geocast routing protocols is given by [MAI04] whereas an in-depth description for C2X communication is detailed in [ETSI11c].

For C2X communication the following message distribution schemes are taken into account:

First of all, the *geographically-scoped broadcast*, which is commonly intended by referring to *geocast*. It is supposed to transmit information, i. e., a C2X message,

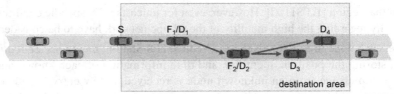

cf. [C2C07]

FIGURE 2.7.: A C2X communication scenario based on the *geographically-scoped broadcast* scheme. The source node S is transmitting a message to all nodes D_j within a specific geographical destination area. This may include multiple intermediate nodes F_i inside the area.

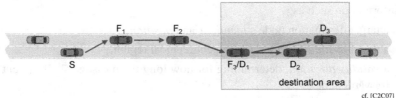

cf. [C2C07]

FIGURE 2.8.: A C2X communication scenario based on the *geographically-scoped broadcast* scheme. The originating source node S is located outside the destination area. Thus, the message is forwarded by multiple intermediate nodes F_i, before it is distribute to all nodes D_j within the destination area.

from a single source station to all stations within a specific destination area. Thereby, the destination area is a geographical region such as detailed in Section 2.5.1. Hence, each ITS station within the destination area, except the source itself, is a destination. Additionally, receiving nodes may forward the message within the destination area. Such a communication scenario is depicted in Figure 2.7.

In addition to distribution requirements, receiving nodes may also retransmit the message within the destination area as long as its validity duration is not expired. This *Keep-Alive Forwarding* (KAF) is intended to ensure message transmission over longer time-scale. Hence, even vehicles entering the destination area later on will receive the message, too, as long as it is valid.

Within a geographically-scoped broadcast, the source node does not necessarily need to be insight the destination area. Hence, the message is forwarded from the source node to the destination area by intermediate forwarding stations. Finally, it is distributed within the destination area, as depicted in Figure 2.8.

Local dangers, such as, e. g., objects on the road or operating emergency vehicles, are of interest within a specific geographical region. Hence, within C2X communi-

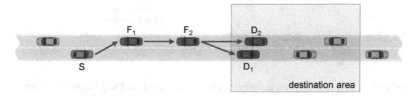

FIGURE 2.9.: A C2X communication scenario based on the *geographically-scoped anycast* scheme. A message is transmitted via multiple intermediate forwarding nodes F_i to any destination nodes D_j inside the destination area and direct communication range of the last forwarding station.

cation the geographically-scoped broadcast scheme is usually applied on DENMs, regarding these local dangers. Such as the DENMs regarded by the Weather Hazard Warning.

A variation of the geographically-scoped broadcast is the *geographically-scoped anycast*. Thereby, the message may be forwarded towards the destination area, but is not distributed inside it. Hence, these messages are only received by nodes within direct communication range of the source node or the last forwarding node outside the destination area, as depicted in Figure 2.9.

In addition to these schemes, a geographical adaption of the unicast, the *geographical unicast*, is defined. It is intended for unidirectional communication from one source station to a single destination station, is defined. Thereby, a message may be transmitted directly, if source and destination are adjacent stations, or forwarded by multiple intermediate stations, otherwise. Therefore, a C2X message contains a destination node's identifier and its last known geographical position. Due to the high mobility, the destination position is regarded as approximated target during determining the forwarding route, only. Additionally, a hop-limit and time information regarding the message validity as well as target's position time stamp may be provided to restrict message delivering efforts. Such a communication scenario is depicted in Figure 2.10. Within C2X communication the geographical unicast scheme may be applied, e. g., on PVDMs, since these messages are intend to be received by a single Roadside ITS Station.

In contrast to other presented message distribution schemes, the *topologically-scoped broadcast* does not rely on geographical information. Instead, the message contains a hop-limit h, thus, each receiving node is decreasing the hop-limit by one and forwards the message as long as

$$h > 0 \qquad\qquad (2.3)$$

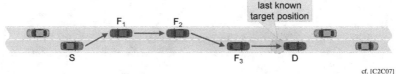

cf. [C2C07]

FIGURE 2.10.: A C2X communication scenario based on the *geographical uni-cast* scheme. The source node S is transmitting a message to the destination node D via multiple intermediate forwarding nodes F_i.

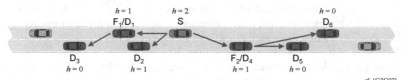

cf. [C2C07]

FIGURE 2.11.: A C2X communication scenario based on the *topologically-scoped broadcast*, with a hop limit of $h = 2$. The message originated by the source S is retransmitted by receiving nodes F_i as long as the hop limit is not reached.

holds. The topologically-scoped broadcast is used for communication from one source to all station within the ad-hoc network and the respective topological scope. Hence, each receiving node is a destination and each, but the leave nodes, is a forwarder. An according communication scenario is depicted in Figure 2.11. Within C2X communication, the topologically-scoped broadcast scheme with a hop-limit of

$$h = 1 \quad , \tag{2.4}$$

is, e. g., applied to CAMs.

Decentralized Congestion Control

Car-to-X communication is based on a wireless network without any central controlling. Thus, messages might collide frequently such that they cannot be decoded by the receiving station.

Hence, a *Carrier Sense Multiple Access with Collision Avoidance* (CSMA/CA) scheme is applied. Thereby, each transceiver listens if there is any communication on the channel and starts a transmission only in the case that the channel is free. Otherwise, if the channel is busy, the transceiver waits for a time period randomly selected out of a specific range before the channel state is checked again.

Nevertheless, collisions will occur, since the channel might be free at listening time but busy when the transmission is started or it might be free at the station's position, but an adjacent station has already started a transmission and the signal does not has reached the sending transceiver, jet. Due to restrictions of wireless communication, it is almost impossible to detect such message collisions.

Since message collisions become more frequently with increasing channel load, *Decentralized Congestion Control* (DCC) is applied to avoid collisions as far as possible. Selected aspects regarding DCC are outlined and discussed in the following.

First of all it has to be regarded that C2X communication aims towards an improvement on road safety. Hence, to ensure a certain *Quality of Service* (QoS), highly safety related messages shall be prioritized. Accordingly, messages include a *traffic class* [ETSI11a] attribute, determining their priority. The actual traffic class depends on the message type, the use-cases, as well as the current scenarios.

To apply both, a prioritization and a reduction of sent messages, within C2X communication CSMA/CA is adjusted accordingly. Hence, the actual range, the waiting time is selected out of if the transceiver detects a busy channel, is determined by the current message's traffic class, the current overall channel load, and the number of tries the transceiver has already started to transmit the current message [ETSI11a].

A complementary approach is based on reducing the transmission power [ETSI11a]. This way, the number of reached and, thus, affected station is reduced. In [TMSH06] such an approach is evaluated, especially regarding *fairness*. However, besides the benefits of this approach, it has to be considered that reducing the transmission power decreases the transmission range and, hence, increases the number of required hops to forward a message. This again increases the total number of messages to be sent.

A more flexible approach on radiation control, the *Secure C2X Beamforming*, is introduced in [SSH09] and particularly detailed in Section A.2. Thereby, targeting on improving security aspects, the method of radiation control by means of beamforming is applied to C2X communication. Thus, an antenna's radiation is not distributed equally into all directions, but focused towards a receiver's position. Regarding DCC aspects, this approach facilitates a reduction of medium usage within unintended directions and, hence, reduces the overall channel usage. However, since, as outlined in Section A.2, Secure C2X Beamforming relies on especial designed antennas, this approach is not applied within current C2X communication designs and standards.

On the other hand, DCC might not only be applied on physical and medium access layer but additionally on higher levels. Hence, e. g., the frequency at which CAMs are generated is reduced [ETSI11b] by adapting the message generation criteria as outlined in Section 2.3.1, if the overall channel load increases. However, this results in less updates such that C2X applications have to rely on outdated data.

Additionally, the message repetition rate, for *forwarding*, and *keep-alive forwarding*, as applied by a message's originator might be reduced. However, it has to be regarded that this adaption results in slower message distribution such that receiving ITS stations may not receive messages in time.

To provide DCC, within C2X communication multiple approaches are combined. Hence, channel load and traffic class adaptive CSMA/CA, transmission power adaption, and message frequency reduction are applied [ETSI11a]. However, as the evaluation in [ACM+13] shows, the actual quality of this combination of different techniques depends highly on the selected parameters, but demonstrates a promising approach.

2.4. Pseudonym Changes

Within Car-to-X communication messages containing vehicle's position, speed, heading, and further meta-data are sent periodically with a frequency of 1 Hz up to 10 Hz as detailed in Section 2.3.1. These messages can easily be linked together to reconstruct the entire path the vehicle has moved [GH05] as outlined in Section A.1.2.

Since, firstly, the number of registered vehicles highly correlates with the number of driving licenses within each household [ADAC11] and, additionally, the total number of driving licenses [BAST07] and total registered vehicles [BK08] correlate, each vehicle may be mapped to almost on person. Consequently, from tracing the movements of a specific vehicle it can be concluded to the movements of a specific person and its identity with high reliability. This opens the goal for countless undesired privacy infringements by unauthorized adversaries and, hence, an immense loss of anonymity. To assure complete *anonymity* a person's identity must not be revealed by any transmitted data [ECK13].

On the other hand, in the case of, e. g., hit-and-run or other criminal acts, determining involved vehicles is desired. Accordingly, a certain level of accountability by means of resolving determined messages to a specific vehicle and, hence, a person's identity by law enforcement agencies is intended.

To provide a compromise between the contrary requirements of anonymity and accountability, a concept based on *pseudonymity*, a weaker form of anonymity, is established within C2X communication [IEEE06]. Thereby, *pseudonyms* are consulted, which are not linkable directly to a vehicle, but may be resolved by respective authorities, if required [ECK13]. However, equipping Vehicle ITS Stations with a single pseudonym does not solve the issues of privacy infringement. Instead, the risk of easy tracing and resolving the driver's identity is still given, as stated above.

An approach based on *pseudonym changes* is commonly accepted as a promising concept to improve privacy and facilitate pseudonymity in C2X communication. Thereby, every VIS possesses multiple pseudonyms and changes them frequently. During a pseudonym change any information, which is included in a message and may serve as an identifier, has to be replaced.

Accordingly, each *pseudonym* has to provide new identifiers for all layers of the C2X communication stack. This includes network address, station ID, IP address, public and private keys with according certificates, etc.

Moreover, even application specific data may be exploited to identify a vehicle and, hence, to map two individual pseudonyms. Especially information with high resolution or combinations of different data, like

- vehicle dimensions with a resolution up to millimeters as used in CAMs [ETSI11b],
- positions and dimensions of single detected situations transmitted in DENM or even sets of situations as proposed in [C2C14],
- multiple positions and probe values as included in PVDM as detailed in Section 2.3.3, or
- traces as used in CAMs [ETSI11b] or within situations in DENMs as proposed in Section 3.2.2,

are extremely individual and, hence, may be easily exploited as unique identifiers. Accordingly, generation of suchlike "identifiers" has to be prevented. Thus, the author of this thesis proposes that

- CAMs may provide vehicle dimensions in decimeters instead of millimeters and the included trace should be reset with every pseudonym change,
- previous detected situations should not be included in DENM after a pseudonym change, as applied in Section 3.2,
- PVDMs have to be encrypted,
- within event detection, as proposed in Section 3.2.2, new messages have to be triggered before the pseudonym change is applied, such that traces included in DENMs do not contain positions gathered before and after the pseudonym change.

Due to the necessity of changing or avoiding identifiers within various instances of a C2X system, the concepts of *Pseudonym Change* has to be deeply integrated within every single layer of the C2X communication stack.

The concept of pseudonymity with pseudonym changes is applied to Car-to-X communication by issuing sets of multiple pseudonyms to ITS stations by central authorities. In order to achieve these *short term identities*, vehicle stations send an according request to the responsible authority. This communication has to be secured by means of encryption and digital signatures. Thereby, a vehicle's *long term identity* is used to authenticate the station towards the pseudonym issuer. Thus, the respective authority may resolve each pseudonym to a corresponding static and unique identity.

A long term identity is preferably assigned on vehicle manufacturing or on vehicle registration. However, due to possible corruption of long term identities, which cannot strictly avoided as outlined in Section A.1.1, interfaces to exchange and revoke identities have to be foreseen. In contrast suchlike interface are not required for short term identities, due to their short validity duration.

However, even if applying the concept of changing pseudonyms is commonly used within field operational tests, e. g., in sim^{TD} [SIMTD10a], preliminary standards [IEEE06] and accepted as one of the promising approaches to ensure a certain level of privacy, tracing cannot completely impede by this technique [SKL$^+$06]. Advocating, e. g., a Kalman filter-based approach to trace vehicles, pseudonym changes may be revealed easily [WMKP10].

Consequently, additional complementary techniques, e. g., *Mix-Zones* [BS04], may be applied to C2X communication systems by means of establishing static Mix-Zones [LBV07] on suitable locations like, e. g., intersections [FRF$^+$07]. Hence, effectiveness of pseudonym changes is increased and improvements to driver's privacy are reached [DDS10]. Moreover, additional enhancements are achieved by forming *Dynamic Mix-Zones* upon vehicle's request [YMM13]. Thereby, a secret group key may be established, e. g., based on geographical cells [SPH11]. Additionally, advanced key agreement schemes, e. g., based on the Diffie-Hellman key exchange [DH76], may be adopted to the C2X communication domain [SÇH11] facilitating dynamic group management, i. e., joining and leaving the group.

2.5. Area and Coordinate Definitions

In order to exchange information about geographical areas and positions between different ITS stations, their shapes and representation has to be agreed on. Hence,

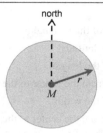

FIGURE 2.12.: The two parameters, i. e., M and r, of a circle as given in a DENM.

within the following sections, regarded geographical areas and their representation, is outlined in the first place.

While coordinates are encoded within the WGS 84 coordinate system between ITS stations and components within a station, for internal representation and calculation the use of a different representation might be selected. Hence, multiple possible coordinate systems and transformations are detailed and discussed subsequently.

2.5.1. Area Definition, Representation, and Encoding

In order to refer a geographical area, the shape of that area is given by either a *circle*, an *ellipse*, or a *rectangle* [ETSI10c].

A circular area is defined by a center point M and a radius r as depicted in Figure 2.12.

To define not only axis-oriented, but also arbitrary oriented rectangles, there are different possible sets of parameters evaluated in field operational tests. In the German research project sim^{TD}, e. g., two points and a width are exploited [SIMTD09j]. Thereby, the points specify two neighbored edges and, therefore, the length and a border of the area. The width is added to the left hand side of that border. However, in current standardization a rectangle is defined by the four parameters as detailed in Table 2.3.

Thereby, M is given as a WGS 84 coordinate, a and b are given in meters, and θ is an angle in degrees, clockwise from north. Such a rectangle, including its parameters, is depicted in Figure 2.13.

Finally, an ellipse is expressed with the same four parameters as used for a rectangle, except that an ellipsoid shape is calculated as depicted in Figure 2.14.

Defining these three shapes in this way provides major advantages. Firstly, the representations may be encoded within a DENM in a very similar manner. Since the four parameters, i. e., M, a, b, and θ, for a rectangle and an ellipse are identical, they,

M	the center of the rectangle (longitude, latitude)
a	the half of the length (distance from the center to the shorter side)
b	the half of the width (distance from the center to the longer side)
θ	the clockwise rotation against north (azimuth angle)

cf. [JH13]

TABLE 2.3.: Summary of the four parameters defining a rectangle within a DENM.

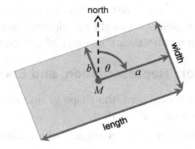

cf. [JH13]

FIGURE 2.13.: The four parameters, i. e., M, a, b, and θ, of a rectangle as given in a DENM.

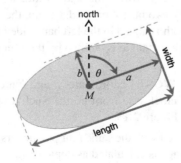

FIGURE 2.14.: The four parameters, i. e., M, a, b, and θ, of an ellipse as given in a DENM.

obviously, may be encoded in exactly the same way. However, even the parameters of the third shape, i. e., M and r, may fit into this representation. Thereby, the center point M is already given, a and b are set equally to the radius r, and θ is set to 0. Secondly, the size of the shape may be changed straightforwardly. Since all shapes are expressed based on the center point in combination with the expansion from that point, they may be magnified or shrunken by just increasing or decreasing a, b, or r, respectively. Furthermore, rotation is applied by only adapting θ. Finally, a Cartesian coordinate system based on the shape may be easily constructed as further detailed in Section 4.3.2.

Within the context of the Weather Hazard Warning application as presented in this thesis, rectangles are exploited to define geographical areas.

2.5.2. Coordinate Definition and Representation

Geographical positions, e. g., provided from a GPS receiver, are commonly given in coordinates according to the *World Geodetic System 1984* (WGS 84) coordinate system. In WGS 84 coordinate system each position is refereed by its two spherical coordinates *longitude* λ and *latitude* φ.[14] As mentioned before, in C2X communication points provided in DENMs are given in such coordinates.

However, for most mathematical operations, it is advantageous not to deal with spherical coordinates but with coordinates in a two dimensional Cartesian coordinate system. Additionally, units within the coordinate system should be interpreted as meters. This simplifies transferring sizes, like distances, lengths, and areas, into the real world. Finally, such a coordinate system can be interpreted as a two dimensional vector space $\mathbb{R}^2$ by using the point's coordinates as components of the point's position vector. Consequently, in the context of this thesis, the conversion of WGS 84 coordinates into a two dimensional Cartesian coordinate system is selected to achieve efficient calculations within a C2X application.

Subsequently, three selected coordinate transformations are detailed and compared.

Universal Transverse Mercator Coordinate System

The *Universal Transverse Mercator* (UTM) coordinate system [DMA89] is a possible choice for such a coordinate system. Positions in UTM coordinates are given by three parameters, the *UTM zone*, an *easting*, and a *northing*.

Thereby, based on the WGS 84 ellipsoid, the world is divided around the equator

---

[14] In fact, there is additionally an *elevation*, i. e., the height above the reference ellipsoid, provided. However, in the context of this thesis, the elevation can be omitted.

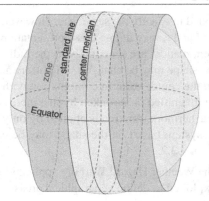

FIGURE 2.15.: An adapted transverse Mercator projection as utilized within the
transformation of WGS 84 coordinates into the UTM coordinate
system. Thereby, the earth's surface is cut by the zone's projection
cylinder.

into 60 *zones*, each a 6° wide band of longitude.[15] The zone numbering increases
eastward, starting at the *International Date Line* with zone 1, which covers 180° to
174° west up to zone 60, which covers 174° to 180° east.

For each zone, a slightly adapted *transverse Mercator projection* is applied. Thereby,
the earth's surface is projected onto a cylinder. To minimize the overall distortion,
the cylinder is cutting the sphere's surface east and west of the center meridian of
the zone.[16] At the equator, these *standard lines* are located ≈ 1.5° to the east and
west, respectively, of the zone's center meridian. This way, the scale factor between
the standard lines is less than 1 with a minimum of 0.9996 at the center meridian of
a zone, at the standard lines the scale factor is 1, and outside it is above 1. Such an
adapted transverse Mercator projection is illustrated in Figure 2.15.

Based on these zones, easting is defined as the distance in meters to the center
Meridian of the respective zone.[17] In contrast, northing is simply defined as the
distance in meters to the equator.[18] Consequently, easting is unambiguous only
within its own zone, while northing is globally unique.

The conversion of a WGS 84 coordinate into the UTM system can not be explained
within a few formulas only. However, the algorithm and related respective equations

---

[15] With the exception of a few non-standard zones.
[16] At a pure transverse Mercator projection the projection cylinder would touch the sphere at the zone's
center meridian.
[17] To avoid negative distances in the left hand side of each zone, an offset of 500 km is added.
[18] Again, to avoid negative distance, 10 000 km are added within the southern hemisphere.

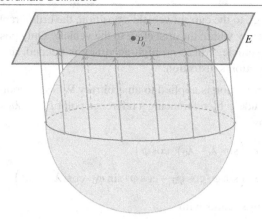

FIGURE 2.16.: The orthographic projection maps points of a sphere onto a flat
projection plane E, which touches the sphere at the reference point
P_0. Thereby, positions located on the opposite side of the sphere as
the reference point are clipped and not mapped to the plane.

are extensively detailed in [DMA89].

To exploit UTM coordinates as a two dimensional Cartesian coordinate system,
easting[19] and northing have to be interpreted as x- and y-coordinates, respectively.
This approach is implemented and tested within the *Road Weather Warning* application in context of the German research project *sim^{TD}* [SIMTD]. However, the
handling of UTM coordinates near the borders of UTM zones and, moreover, of
points positioned in different zones, is rather challenging.

Orthographic Projection

A second approach to achieve a two dimensional Cartesian coordinate system is
to apply an *orthographic projection*. Thereby, the earth's surface is projected onto
a tangent projection plane E, where an intuitive Cartesian coordinate system is
given. The projection plane touches the surface of the earth, which is assumed to
be a perfect sphere, at a reference position P_0. Such an orthographic projection is
depicted in Figure 2.16.

Due to the projection of a spherical surface onto a flat plane, the distortion, in terms
of shrunk distances, growths with an increasing distance to the reference point.
Finally, at a distance of 90° the distortion reaches its maximum and no projection
of points placed beyond this limit is possible anymore. However, if the reference

---

[19] For easting, the current UTM zone has to be regarded, in addition.

position P_0 is set, e. g., to the current vehicle's position, the orthographic projection provides a high accuracy within the area close to the vehicle. Only positions located far away from the vehicle, where no high accuracy is needed, are significantly affected by the projection's distortion.

The orthographic projection is applied to an arbitrary WGS 84 coordinate P with longitude λ and latitude φ by referencing a reference point $P_0 = (\lambda_0, \varphi_0)$ according to, e. g., [SNY87], by

$$x = r \cdot \left(\sin(\lambda - \lambda_0) \cdot \cos \varphi \right) \tag{2.5}$$

$$y = r \cdot \left(\sin \varphi \cdot \cos \varphi_0 - \cos \varphi \cdot \sin \varphi_0 \cdot \cos(\lambda - \lambda_0) \right) \quad , \tag{2.6}$$

whereas r is the earth's radius with

$$r = 6371000.8 \, \text{m} \tag{2.7}$$

and x and y are the Cartesian coordinates of P. Thereby, the point P is located outside the projection plane and no projection is possible if

$$0 > \arccos \left(\sin \varphi \cdot \sin \varphi_0 + \cos \varphi \cdot \cos \varphi_0 \cdot \cos(\lambda - \lambda_0) \right) \tag{2.8}$$

holds.

The projection of a Cartesian coordinate back to its WGS 84 representation is achieved by applying

$$\lambda = \lambda_0 + \arctan \left(\frac{x \cdot \sin \varsigma}{\rho \cdot \cos \varphi_0 \cdot \cos \varsigma - y \cdot \sin \varphi_0 \cdot \sin \varsigma} \right) \tag{2.9}$$

$$\varphi = \arcsin \left(\sin \varphi_0 \cdot \cos \varsigma + \frac{y \cdot \cos \varphi_0 \cdot \sin \varsigma}{\rho} \right) \tag{2.10}$$

with

$$\rho = \sqrt{x^2 + y^2} \tag{2.11}$$

$$\varsigma = \arcsin \frac{\rho}{r} \quad . \tag{2.12}$$

Delta-Lambda-Equator Coordinates

A third possible coordinate transformation again builds up the coordinate system based on a reference position P_0. Thus, no reference to a respective zone is needed. Thereby, Cartesian coordinates are achieved by taking the distance in meters from the reference point's longitude λ_0 (the *reference meridian*) Δ_λ as x-coordinate and

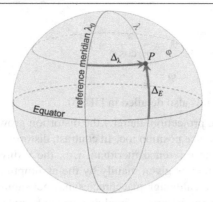

FIGURE 2.17.: Within the DLE coordinate system the x- and y-coordinates of a point are given by the points distance to a reference meridian Δ_λ and distance to the equator Δ_E, respectively.

the distance in meters to the Equator Δ_E as y-coordinate, respectively. For a point P this approach is depicted in Figure 2.17.

The basic idea of this approach was firstly proposed by the author of this thesis in [JH13]. Thereby, the reference meridian was set to the *International Reference Meridian*, such that $\lambda_0 = 0°$ holds. In contrast to this static reference meridian, for the novel *Delta-Lambda-Equator* coordinates (DLE) the reference meridian might be selected dynamically based on the current scenario.

For simplification, within the coordinate transformation, the earth again is assumed to be a perfect sphere. The radius r of this sphere is set to the mean radius of the WGS 84 ellipsoid, i.e.,

$$r = 6\,371\,000.8\,\text{m} \quad . \tag{2.13}$$

An arbitrary WGS 84 coordinate with longitude λ and latitude φ is converted into a point $P = (x, y)$ regarding a reference meridian λ_0 by utilizing the arc angle[20] such that

$$x = r \cdot (\lambda - \lambda_0) \cdot \cos\varphi \tag{2.14}$$

$$y = r \cdot \varphi \quad . \tag{2.15}$$

Such a point P of the Cartesian coordinate system is converted back into a WGS 84

---

[20] In contrast, in [JH13] the respective x- and y-coordinates are determined based on the calculation of the orthodromic distance. Thereby, the equation was simplified by $\sin(0) = 0$ and $\cos(0) = 1$. However, by using the arc angle, the calculation is further simplified significantly.

coordinate by applying

$$\lambda = \frac{x}{r \cdot \cos \varphi} + \lambda_0 \qquad (2.16)$$

$$\varphi = \frac{y}{r} \qquad . \qquad (2.17)$$

This advanced approach is also detailed in [JH15].

As within orthographic projection, thereby the distortion growths in relation with the distance to the reference position too. In contrast, distortion growths in relation with the distance to the reference meridian, i. e., the x-direction only. Along the y-direction inaccuracy is given mainly by the assumption that the earth is a perfect sphere. Moreover, although inaccuracies may get significant large, even for x-direction, there is no restricting maximal distance to be regarded.

However, within several kilometers around the reference point no noticeable distortion has to be considered. Since for Car-to-X applications the near area around the vehicle is regarded, the reference point may be set to, e. g., the vehicles current position, to achieve highest transformation accuracy for relevant areas.

In order to prove the usability of DLE coordinates, this concept was implemented and extensively evaluated within the *Weather Warning* application in context of the research project DRIVE C2X [DRIVE] as further detailed in Section 6.4 and Section 7.1.

Argumentative Comparison and Recommendation

Comparing the presented approaches, it is obvious that all three coordinate systems, of course, cannot be used in a global manner, i. e., handling coordinates that are far apart from each other. Nevertheless, in local scale of multiple kilometers, as needed for C2X applications, all three are highly appropriate.

However, conversion of WGS 84 into UTM coordinates and vice versa needs extensive computations[21], while for applying an orthographic projection or conversion into the DLE coordinates just the respective enumerated simple equations have to be applied. Since computation resources are highly restricted in automotive domain, the high transformation efficiency is a huge advantage of the proposed transformation into DLE coordinates. In addition, relative positioning and distances in local scale do not suffer from inaccuracies based on simplifying the earth as a sphere. Hence, a high accuracy for Car-to-X communication purposes is ensured.

Consequently, the author of this thesis recommends to use the outlined DLE coordinates to convert WGS 84 coordinates into a two dimensional Cartesian coordinate

---

[21] The related formulas exceed the scope of this thesis, but are detailed in [DMA89].

system. For the context of this thesis, it is accordingly assumed that points are given in or converted into the proposed DLE coordinates.

Regarding Coordinates within the Vector Space $\mathbb{R}^2$

As mentioned, for some mathematical calculations positions may be regarded as points not only within a two dimensional Cartesian coordinate system, but within a two dimensional vector space $\mathbb{R}^2$. For all three detailed coordinate systems, a transformation into the vector space is possible.

Within $\mathbb{R}^2$ for each point P, there is the *position vector* $\vec{P}$. A position vector $\vec{P}$ for the point $P = (x, y)$ is defined as

$$\vec{P} = \begin{pmatrix} p_1 \\ p_2 \end{pmatrix} = \begin{pmatrix} x \\ y \end{pmatrix} \quad . \tag{2.18}$$

Hence, due to the current needs, each position may be regarded as point with x- and y-coordinate in context of a two-dimensional Cartesian coordinate system or as point within the vector space $\mathbb{R}^2$ represented by the according position vector. Thus, calculations as detailed in Sections 3.2.3, 4.3.1, 4.3.2, 4.3.3, and 5.1, respectively, are partially performed within a vector space.

3. Weather Hazard Detection

Modern vehicles contain a multitude of sensors. Several of them, e. g., outside temperature, rain intensity, or brightness probes, are suitable for detecting the current weather condition. These sensors may be further complemented by observing driver's actions, e. g., activation of low beam lights, selected level of front shield wipers, or state of rear fog light. All this information is observed and combined together to detect hazardous weather situations by means of the Weather Hazard Warning application within vehicles.

Firstly, available vehicle data have to be collected and converted in a suitable unified representation as detailed in Section 3.1. Subsequently, current evidences on outside weather conditions have to be evaluated. This detection process is detailed in Section 3.2.

In addition to the mentioned detection within *Vehicle ITS Stations* (VIS), supplementary weather situations are detected by Central ITS Stations as further described in Section 3.3. These infrastructure-based detection takes into account not only gathered vehicle data but, additionally, complementary and highly reliable measurements from, e. g., weather measurement stations placed along the roads.

3.1. Vehicle Data Processing

In the first step to detect hazardous weather conditions within vehicles, vehicle probe data has to be collected. Within current vehicles, the vast majority of sensor measurements are distributed via one or multiple CAN buses [ISO03a].

Thereby, if there are multiple buses, typically some of them are used especially for time critical active safety systems, e. g., an *Electronic Stability Program* (ESP) or an *Anti-lock Braking System* (ABS), which depend on hard real time constrains. In contrast to such a *High-speed CAN* [ISO03b], information about low beam lights, turn signals state, and outside temperature are bound to a *Low-speed CAN* [ISO06]. Thus, communication of time critical information is not interfered by such comfort or management data. Accordingly, essential vehicle data are distributed within multiple buses. Hence, in general, Car-to-X applications have to be placed on a unit connected to all different CAN buses installed in a vehicle.

3.1.1. Vehicle Data Unification

Availability of vehicle data within the CAN varies not only with vehicle's equipment, but also between vehicle models, issuing date, and, by nature, most heavily between vehicle manufacturers. Moreover, data representation is highly inhomogeneous between different vehicles. Hence, e. g., the current vehicle speed may be expressed in meters per second (m/s) in contrast to kilometers per hour (km/h) or the activation of the turn signal switch is indicated on the CAN in contrast to the frequently recurring activation of the turn light bulb. Furthermore, encoding on byte- and bit-level is more or less arbitrarily.

Consequently, in order to make applications independent from the actual vehicle platform, a C2X application may not use CAN data directly. Instead, a special component for proving vehicle data has to be introduced. This component has to be adjusted or configured to the particular equipment of the current vehicle and to the especial data representation within the current CAN.

Such a component is described, e. g., in [EK06], or used within multiple C2X communication field operational tests [PREDR09] [SIMTD09g] [DRIVE11c]. Within this *Vehicle Data Provider* (VDP) system component the current vehicle's CAN data are converted into a precisely specified set of basic and enhanced common data objects with unified representation. Thus, each C2X system component and each C2X application may obtain comparable vehicle information provided in determined frequencies of either 1 Hz, 2 Hz, 5 Hz, or 10 Hz. The respective frequency depends on how frequently the data change. Consequently, applied algorithms and entire applications are comparable and highly interchangeable between different vehicles even from multiple manufacturers.

Although C2X applications are developed independently by multiple manufacturers, they have to work together cooperatively. To ensure similar behavior of data providing components in different C2X communication units a basic set of mandatory vehicle data is defined in the *Common Data Dictionary* (CDD) [ETSI13b]. This way, an almost unified and comparable interface to vehicle data is intended on which system components and applications rely on.

For the Weather Hazard Warning application, consulted data elements, according to [ETSI13b], are enumerated in Table 3.1.[22] However, the ETSI specification

---

[22] In fact, the data elements usually are composed out of further sub-elements, e. g., a *-confidence* and a *-value* element. This complexity is omitted for reasons of simplification in Table 3.1, Table 3.2, and

Data Element	Description	Unit
Curvature	curvature of actual vehicle trajectory	m^{-1}
ExteriorLights	state vector of the exterior light switches (only lowBeamHeadlightsOn, fogLightOn)	boolean
LongitudinalAcceleration	vehicle acceleration at longitudinal direction	m/s^2
Speed	speed of a moving vehicle	m/s
Temperature	outside air temperature	°C
VerticalAcceleration	vehicle acceleration at vertical direction	m/s^2
YawRate	vehicle rotation around its center	°/s

TABLE 3.1.: Summary of data elements according to ETSI consulted in the Weather Hazard Warning application.

Data Element	Description	Unit
AntilockBrakingSystem	state of the ABS	boolean
BrakingPedal	level the braking pedal is pushed	%
ElectronicStabilityProgram	state of the ESP	boolean
ExteriorLights	state vector of the exterior light switches (add frontFogLightOn)	boolean
GasPedal	level the acceleration pedal is pushed	%
RainSensor	intensity of rain- or snowfall	mm/h
TractionControlSystem	state of the TCS	boolean
WiperSystem	rain wiper speed	wipes/min

TABLE 3.2.: Summary of additional data elements consulted in the Weather Hazard Warning application but not included in current version of ETSI specification.

[ETSI13b] is intended to be extended during time. To improve detection capabilities and ensure sufficient detection reliability, the Weather Hazard Warning application relies on additional data elements as preliminary proposed in Table 3.2. Such data elements are intended to be added to [ETSI13b] in the near future and, hence, applied within the context of this thesis.

Upon such a common basis the standardized definition of C2X application behavior is facilitated.

Table 3.3. Furthermore, some values are provided with a scaling factor, which is omitted for reasons of simplification, too.

Data Element	Description	Unit
Latitude	absolute geographical latitude	°
Longitude	absolute geographical longitude	°
Heading	direction of driving, relative to geographical north	°

TABLE 3.3.: Summary of data elements related to geographical positioning information according ETSI specification.

3.1.2. Improved Positioning

Additionally to the previous mentioned vehicle specific data, C2X applications of
course rely on the current vehicle's position. The position is provided by a *Global
Navigation Satellite System* (GNSS), usually the *Global Positioning System* (GPS)[23]
but also others like, e. g., the European *Galileo* are feasible.

In addition to the position, the GPS signal contains highly precise timing information. This timing information may be used for synchronization of system time.
Moreover, if multiple vehicles rely on this timing information, the system times are
inherently synchronized between individual stations.

Each position is given according to the *World Geodetic System 1984* (WGS 84)
ellipsoid [NIMA00]. It mainly consist of a *latitude* value in the range of $-90.0°$
to $90.0°$, a *longitude* value in the range of $-180.0°$ to $180.0°$ and an elevation in
meters above the reference ellipsoid. However, elevation is not considered in the
Weather Hazard Warning application.

Just like the vehicle specific data, representation of positioning information is
standardized for cooperative usage in C2X communication [ETSI13b] and may be
provided by the *Vehicle Data Provider* component as introduced in Section 3.1.1 or
a separate component as suggested in [ETSI09].

But, in addition to the named geographical positioning information, a direction of
driving is required, which may be calculated from prior movements and usually, is
provided by a common GPS receiver. Accordingly, the data elements as enumerated
in Table 3.3 are considered within the Weather Hazard Warning application too.

Due to multiple interferences, reflections, and other measuring inaccuracy the
nominal GPS accuracy is about $10-15$ meters. To improve location accuracy,
Differential GPS (DGPS) is introduced. Thereby, additional correction information is broadcasted by local short range transmitters, cellular technologies, or, in

---

[23] Full name: Navigation System with Timing and Ranging – Global Positioning System (NAVSTAR-
GPS)

larger scale, by satellites as, e. g., within the *European Geostationary Navigation Overlay System* (EGNOS) [ESA06]. Combining GPS positioning with correction information a positioning accuracy up to about 1 meter is reachable in case of ideal conditions and the best implementations [ZOG13].

However, in highly dynamic environments, as within moving vehicles, this precision is hard to achieve. Additionally, satellite signals may be shielded by building or mountains as well as within tunnels. In order to provide essential position information, *dead reckoning* is consulted in C2X communication stations to improve DGPS positioning and even to overcome short periods of GPS signal losses.

Thereby, local vehicle movement related data, like vehicle speed, yaw rate, acceleration, steering wheel angle, etc., are observed over time to estimate the current vehicles position based on a previously known position. Usually, a *Kalman filter* is advocated to improve vehicles positioning [RS07] [MSS10] [SCH10]. Such techniques are effective even with low cost GPS receivers [HMS03]. Moreover, the benefit for C2X application is demonstrated in [KLA08] by the example of the *Intersection Collision Warning* application.

Consequently, DGPS and a component for *Improved Positioning* is integrated within the field operational test system architecture of the C2X research project sim^{TD} [SIMTD09g] and is suggested in the C2X application definition [ETSI09].

3.1.3. Weather Profiles

As introduced in Section 1.1, within the *Weather Hazard Warning* application, vehicles can be regarded as mobile sensors gathering probe values along their way on the road.[24] These additional measurements are utilized to enhance information gained from conventional sources, e. g., weather measurement stations.

For situation detection within vehicles, vehicle data may simply be gathered by a direct connection to the vehicle internal CAN. To facilitate detection within *Central ITS Stations*, vehicle data have to be gathered, stored, and transmitted within appropriate messages. Hence, within defined intervals samples of vehicle sensor data are captured and stored in combination with referencing information, i. e., a time stamp and the vehicle's current position. In order to transmit these recorded samples, the data are packed into a suitable *Probe Vehicle Data Message* (PVDM) message, e. g., as detailed in Section 2.3.3.

---

[24] This concept is not limited to the *Weather Hazard Warning* only, but also realize within other C2X use-cases, especially those, who depend their calculations on the current traffic situation, like, e. g., *Traffic information and recommended itinerary* or *Enhanced route guidance and navigation*.

Such a *weather profile* consists of probe vehicle data gathered by a vehicle along its way on the road. Thereby, independent on the current expression, exploited unit, encoding, or resolution, each sample of vehicle sensor data within a weather profile may consists of a

- time stamp,
- position (longitude, latitude),
- outside air temperature,
- rain wiper speed,
- rain sensor level,
- rear and front fog lights state,
- low beam headlights state,
- atmospheric pressure, and
- relative humidity,

if the related sensor is available within the vehicle.

The profiles contain unprocessed raw values. To achieve reliable information they have to be extensively processed. However, it has to be assumed that processing capabilities within *Vehicle ITS Stations* will be hard limited. Consequently, PVDMs are intended to be sent to *Central ITS Stations*, only. These stations have access to conventional data sources and typically own large processing capabilities. This way, CIS may be able to achieve a reliable overview about the local and large-scale weather conditions. Hence, advanced warning messages and forecasts are facilitated.

To transfer PVDMs to central stations, *Roadside ITS Stations* are exploited as intermediate forwarders. However, only several RIS may be connected to a CIS intended to process weather profiles. Consequently, *Service Announcement Messages* (SAM) are defined [ETSI14]. Roadside station capable to receive and forward PVDMs may broadcast respective SAMs. Thus, passing vehicles may trigger transmission of weather profiles within the presence of the roadside station.

However, C2X messages are transmitted over the air, an inherently unsecured medium, such that typically broadcasted messages are received by every ITS station within reception range and message destination area. Thus, message's content is exposed to every receiving station. Since PVDMs contain collected personal mobility data, e. g., position, speed, etc., providing these data to every ITS station may be a huge threat to driver's privacy. Consequently, the content of PVDMs shall be secured by means of encryption [IEEE06], such that personal data may only be visible to receiving trustworthy Central ITS Stations.

3.2. Situation Recognition

Current vehicles contain a various set of highly sophisticated sensors. These measurements in combination with driver's action provide sufficient evidence to determine the current weather situation outside of the vehicle. Hence, hazardous weather situations may be detected by the Weather Hazard Warning application within vehicles.

In early deployment phases, C2X communication systems will be integrated into existing vehicle architectures. These architectures are not intended to provide every measurement and information to all systems. Hence, e. g., a rain sensor may be directly connected to the wiper controlling system and, thus, measurements are available there only. As a consequence, few data from vehicle sensors or information about driver's actions may be available to C2X applications at the beginning.

Accordingly, a basic situation detection, relying only on few and commonly available data, is proposed by the author of this thesis in Section 3.2.1 in order to facilitate weather detection even with a small set of simple sensors. Based upon this basic detection algorithm, a specification for a simple "day-one" Weather Hazard Warning application will be introduced.

Once C2X communication systems are established and highly integrated into vehicle architectures, various sensor data will be available as well as computation systems facilitating complex calculations. Hence, an advanced situation detection algorithm is suggested in Section 3.2.2.

3.2.1. Basic Situation Detection

This section details a basic weather situation detection algorithm. It shall be implemented in early Weather Hazard Warning applications to overcome the gap where only few sensor data are available to C2X communication systems. The detailed rudimentary algorithm is developed to serve as basis for the ongoing development of a unified simple application specification [C2C14][25].

The basic Weather Hazard Warning application is capable to detect three different kinds of weather events. The underlying detection algorithm is identical for all three situation types, i. e., rain, fog, and slippery roads.

The detection is triggered periodically after a use-case specific *repetition time* t_{rep}. In a first step it has to be verified if the use-case specific trigger conditions are

---

[25] Within this specification the application is referred to as *Adverse Weather Warning*.

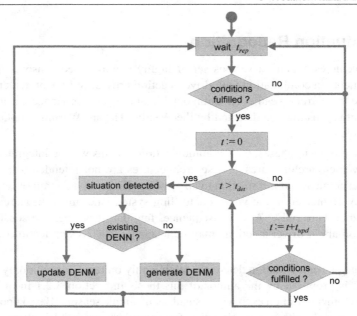

FIGURE 3.1.: Visualization of the basic situation detection algorithm, shared for all
            weather event types supported by the basic Weather Hazard Warning
            application.

fulfilled. To avoid false-positive detections, these conditions have to be met for a
use-case specific *detection time* t_{det}. Hence, the conditions are frequently verified
within a use-case specific *update time* t_{upd}. If the trigger conditions are not fulfilled
in the first step or not met during the whole detection time, the detection is aborted.
Otherwise, if they are met, a situation is detected and an according DENM is
generated or, if a still valid message is transmitted by the vehicle, this message is
updated. This common basic detection algorithm is depicted in Figure 3.1.

To ensure reliable detection, the common basic detection algorithm has to be
repeated, e. g., every 20 s in all use-cases. Since detection time may vary depending
on the actual use-case and the detection algorithm may be aborted after a certain
time during execution, the repetition time has to be adapted accordingly. Hence,
the elapsed detection time t has to be regarded by applying

$$t_{rep} = 20\,\text{s} - t \qquad (3.1)$$

in all algorithm iterations, respectively.

Dense Fog					
Trigger Condition			t_{det}	t_{upd}	**Quality**
	rear fog light	activated	20 s	1 s	1
and	low beam	activated			
and	vehicle speed	≤ 22 m/s			($\approx$ 80 km/h)
	same as above		20 s	1 s	2
but	vehicle speed	≤ 16 m/s			($\approx$ 60 km/h)

TABLE 3.4.: Summary of trigger conditions to detect dense fog situations, including detection time t_{det}, update time t_{upd} and associated information quality.

Based on a reduced set of available sensor data, use-case specific trigger conditions have to be fulfilled to indicate the presence of a weather event. According to the consulted sensors, a specific *information quality* is reached and included into the respective DENM. The information quality gives evidence on the reliability of the detection.

The basic Weather Hazard Warning application is capable to detect either *dense fog*, *heavy rain*, or *slippery road* events. A dense fog situation may be detected by mainly regarding the activation of the rear fog light. These trigger conditions as well as associated information, i. e., information quality, detection timed, and update times, are summarized in Table 3.4. Heavy rain situations may be mainly indicated by observing the front shield wiper level. Additionally, a rain sensor may be regarded if available to increase detection reliability. Appropriate complete trigger conditions as well as associated information are summarized in Table 3.5. Finally, a slippery road situation may be recognized at accelerating or decelerating. In both cases, outside temperature is considered. In combination with the throttle, vehicle acceleration[26], and activation of the *Traction Control System* (TCS) a situation is indicated during acceleration. According trigger conditions as well as associated information are summarized in Table 3.6. In contrast, during braking maneuvers, braking pedal, vehicle deceleration[27] and activation of the ABS is consulted. Respective trigger conditions as well as associated information quality, detection times, and update times are summarized in Table 3.7.

After a situation is indicated by the basic detection algorithm, it has to be verified if the vehicle already transmits a valid C2X message of this use-case. Obviously,

---

[26] By experimental results, the typical vehicle acceleration of 3 m/s^2 on try asphalt is determined. On snow covered roads a typical acceleration of below 1.5 m/s^2 is reached [BNM12].

[27] Hard braking vehicles may reach a deceleration of up to about 7 m/s^2 on dry asphalt [FHWA09]. Decreased deceleration of below 3 m/s^2 on snow and ice is evaluated, e. g., in [LU07].

Heavy Rain					
Trigger Condition			t_{det}	t_{upd}	**Quality**
	wiper level	maximum	20 s	1 s	1
and	low beam	activated			
and	vehicle speed	≤ 28 m/s			(≈ 100 km/h)
	same as above		20 s	1 s	2
but	vehicle speed	≤ 22 m/s			(≈ 80 km/h)
	wiper level	maximum	20 s	1 s	3
and	rain sensor level	$\geq 90\%$			
and	low beam	activated			
and	vehicle speed	≤ 28 m/s			(≈ 100 km/h)
	same as above		20 s	1 s	4
but	vehicle speed	≤ 22 m/s			(≈ 80 km/h)

TABLE 3.5.: Summary of trigger conditions to detect heavy rain situations, including detection time t_{det}, update time t_{upd} and associated information quality.

if there exists no such message, transmission of a new DENM shall be triggered. Accordingly, if there is a message, it has to be updated with new information. Thereby, the event identifier *ActionID*, once assigned during initial DENM creation, stays unchanged during all updates.

The situation's position and detection time is set according to the vehicle's geographical position and beginning of the respective detection, i. e., position and time at start of the according algorithm iteration. Additionally, the direction of driving is included into the DENM. Hence, based upon the delivered position, receiving vehicles may determine towards which direction the situation is spread.

The situations validity duration within the message's *Management Container* is set for new messages or reset for updated messages to a common validity time of, e. g., 300 s, respectively. Thereby, the reliable message distribution is ensured by a message repetition rate of, e. g., 4 s and a relevance area of, e. g., 5000 m.

In the case of DENM updates former event information may be kept within the DENM additionally. Hence, previous situation position and detection time is attached to a *weather history* within the *Alacarte Container* of the updated DENM. Thereby, it has to be ensured that the history contains only elements whose detection time is not older than the common validity duration.

This way, it is ensured that current information about the weather situation is

Slippery Road (at acceleration)					
Trigger Condition			t_{det}	t_{upd}	**Quality**
	TCS	active	300 ms	100 ms	1
and	throttle	$\leq 30\%$			
and	outside temperature	$\leq 2\,°C$			
and	vehicle speed	$\leq 8\,m/s$			(≈ 30 km/h)
	TCS	active	200 ms	100 ms	2
and	throttle	$\leq 30\%$			
and	vehicle acceleration	$\leq 1.5\,m/s^2$			
and	outside temperature	$\leq 2\,°C$			
and	vehicle speed	$\leq 8\,m/s$			(≈ 30 km/h)
	same as above		200 ms	100 ms	3
but	vehicle acceleration	$\leq 1.0\,m/s^2$			
	same as above		200 ms	100 ms	4
but	vehicle acceleration	$\leq 0.7\,m/s^2$			

TABLE 3.6.: Summary of trigger conditions to detect slippery road situations at acceleration, including detection time t_{det}, update time t_{upd} and associated information quality.

distributed. At the same time, loss of previous information is prevented without flooding the communication channel with a huge amount of single messages.

3.2.2. Advanced Situation Detection

This section details an advanced weather situation detection algorithm. A draft idea of this approach is outlined in [JH11], which is refined in improved within the following. It may be utilized in future implementations of the Weather Hazard Warning application, when an extended range of sensor data and increased calculation resources are available to C2X communication systems, eventually.

Situation Detection

The advanced detection, as a novel approach, provides continuous evaluation of the current sensor data in combination with enhanced fault tolerances. To express the location of the weather situation most precisely, the affected area is encoded as a rectangle around the path the vehicle is driven while detecting the situation. This

Slippery Road (at deceleration)					
Trigger Condition			t_{det}	t_{upd}	**Quality**
	ABS	active	300 ms	100 ms	1
and	braking pedal	$\leq 40\%$			
and	outside temperature	$\leq 2\,°C$			
and	vehicle speed	$\leq 14\,m/s$			($\approx 50\,km/h$)
	ABS	active	200 ms	100 ms	2
and	braking pedal	$\geq 40\%$			
and	vehicle deceleration	$\leq 3.0\,m/s^2$			
and	outside temperature	$\leq 2\,°C$			
and	vehicle speed	$\leq 14\,m/s$			($\approx 50\,km/h$)
	same as above		200 ms	100 ms	3
but	vehicle deceleration	$\leq 2.0\,m/s^2$			
	same as above		200 ms	100 ms	4
but	vehicle deceleration	$\leq 1.2\,m/s^2$			

TABLE 3.7.: Summary of trigger conditions to detect slippery road situations at deceleration, including detection time t_{det}, update time t_{upd} and associated information quality.

geometric shape is determined efficient by just a few information[28] while providing high adaption of the road's course[29].

The proposed enhanced situation detection is capable to distinguish five different types of weather events. For all different weather types, the detection of situations is based upon an identical underlying detection algorithm.

To improve fault tolerance against temporarily alternating vehicle data, a *detection state s* is maintained while repeating the algorithm. Hence, if a weather event indicating vehicle data toggles for a short time, this jitter most likely does not interfere the situation detection. This way, false event detection based on short appearance of runaway values or accidental pushed buttons, can be avoided. According inertness is applied for detecting the end of an event. The different detection states are enumerated and detailed in Table 3.8. Obviously, the detection is at *null*-state on system start-up.

To evaluate the current weather condition, the advanced detection algorithm peri-

---

[28] The definition of a rectangle and its parameters is detailed in Section 2.5.1.
[29] In Section 7.3 the accuracy of determined rectangles is evaluated.

State	Description
null	The default state. This state is applied, if no event indicating vehicle data are observed.
observing	This state is exploited to indicate that vehicle data, which may prove a weather event, are received for a short time.
detected	Within this state, a weather event is confirmed. Hence, periodically, as well as on entering and leaving the state, updated DENMs are triggered.

TABLE 3.8.: Description of the different detection states adopted by the advanced weather situation detection.

odically verifies use-case specific trigger conditions at a high rate. Hence, a short update time t_{upd} of, e. g.,

$$t_{upd} = 100\,\text{ms} \quad , \tag{3.2}$$

is applied.

Within each such iteration, the use-case specific trigger conditions are evaluated directly at the beginning. If the conditions are evaluated to be false and the detection algorithm is in its initial state, i. e., *null*, there is nothing to do. Hence, the state is left unchanged and the current iteration ends, immediately.

However, if the conditions are verified, a trace-point is generated and appended to a local trace, i. e., the local driven path history. Subsequently, it has to be verified if the vehicle has already entered a weather event, i. e., detection state is set to *detected*. If not, the triggering conditions may be evaluated as true for the first time. Thus, the *observing*-state is not entered before. In this case, since the conditions may be met just for a short time, the state is set preliminarily to *observing*, the timer t, reporting the time elapsed since the conditions are met, is started, and the current iteration ends.

In contrast, if the *observing*-state was already set, it has to be proved if the conditions are met long enough to surely indicate a weather situation. This is assumed if the vehicle data indicate the weather conditions for at least a use-case specific detection time t_{det} of, e. g.,

$$t_{det} = 20\,\text{s} \quad . \tag{3.3}$$

Hence, it has to be evaluated if t is greater than the detection time

$$t > t_{det} \quad . \tag{3.4}$$

If not, the timer is increased by the elapsed time and the current iteration ends.

Otherwise, a new weather event is detected. Accordingly, the detection state is set to *detected* and an initial DENM, containing detailed information about the situation, is triggered. As after each generation of a DENM, a respective counter t_{send}, indicating the elapsed time since last sent DENM is reset and the so far recorded trace is erased before the current iteration ends.

However, if the current state was already set to *detected* after the trace-point was generated, i. e., a weather event was confirmed during previous iterations, a periodically update of the former sent DENM may be necessary. In doing so, it is ensured that updated DENMs are provided within a common repetition time t_{rep} of, e. g.,

$$t_{rep} = 30\,\text{s} \quad . \tag{3.5}$$

Thereby, the time elapsed since the last DENM was sent, indicated by t_{send}, is evaluated to be greater that this repetition time

$$t_{send} > t_{rep} \quad . \tag{3.6}$$

In that case, the current iteration ends with the triggering of a new DENM and respective reset of t_{send} and the recorded trace. Otherwise, the iteration ends by increasing t_{send} by the time elapsed since the last iteration.

On the other hand, the triggering conditions may not be met in the first step, but the detection state is not set to the default state, i. e., either *detected* or *observing*. This only occurs because of temporary glitches in the sensor data. Hence, the state is not left immediately, but only if the conditions are not met for at least a minimum event leaving time t_{exit} of, e. g.,

$$t_{exit} = 5\,\text{s} \quad . \tag{3.7}$$

Accordingly, the time elapsed within the triggering conditions are not met, indicated by t_{void}, is compared to this threshold. Thus, as long as

$$t_{void} > t_{exit} \tag{3.8}$$

does not hold, t_{void} is increased by the time elapsed since the last iteration. Otherwise, i. e., the triggering conditions are not verified for a longer time, the hazardous weather event is assumed to be finished. Consequently, the detection state is set again to the default state *null*, the timer t_{void} is reset, and the current iteration ends with according reset after triggering of a respective final DENM.

The entire advanced detection algorithm consulted for detecting all weather types within the enhanced Weather Hazard Warning application is visualized in Figure 3.2.

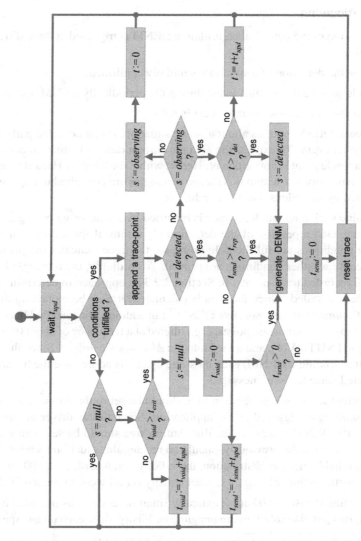

FIGURE 3.2.: Visualization of the advanced situation detection algorithm, shared for all weather event types supported by the Weather Hazard Warning application.

Message Triggering

During iteration of the detection algorithm, a DENM is triggered at three different points in time.

1. When the detection of a situation is initially confirmed.
2. While passing the weather event during the periodically DENM updates.
3. When the hazardous weather event is left.

Thereby, each DENM notifies about a single situation, i. e., a bounded part of the whole weather event. To express the area the current detected situation occupies, reporting a single point is not sufficient. Hence, within the Weather Hazard Warning application as proposed within this thesis, rectangles are consulted to express the area affected by hazardous weather conditions.

To determine such a rectangle, a trace is recorded by means of gathering a trace-point during each repetition of the detection algorithm, if the use-case specific triggering conditions are fulfilled. In doing so, the trace indicates the positions where specific weather conditions are proved. Accordingly, the rectangle has to include all recorded trace-points. In Section 3.2.3 an approach to determine such a rectangle is detailed. Once the area is calculated, it may be included into the *Location Container* of the respective DENM[30]. In addition, the recorded trace may be reduced to a few significant points, e. g., filtered at a frequency of about 0.5 Hz or according to [NHTSA11], and added into the *Alacarte Container*. Since the trace is reset after sending a DENM, each message update is addressing the hazardous area detected since last sent message.

The respective situation's generation time should be set to the moment when the message sending is triggered by the application. Since timely driver notification requires early DENM reception, the distribution area should be set to an area of, e. g., 5000 m around the previously calculated rectangular situation location[31]. To facilitate a reliable message distribution, the validity duration of, e. g., 300 s and the message repetition time of, e. g., 4 s, is set equally for all weather related DENMs.

When selecting the *ActionID* at periodically situation updates as intended by the algorithm, two possible solutions are imaginable. Firstly, since each update specifies

---

[30] Although, the current version of the DENM specification [ETSI13c] does only support definition of circular area of some predefined sizes, within context of this thesis an upcoming enlargement of this sparse possibility to specify a geographical region is assumed. Accordingly, the Weather Hazard Warning is proposed in accordance to the presumed changes.

[31] As detailed in Section 2.5.1, such an enlargement only requires increasing the parameters a and b by 5000, respectively.

a new independent affected area, i. e., the hazardous area since last sent message, it may be interpreted as individual situation. Hence, each update represents a new independent DENM with own ActionID. Thus, previous DENMs stay unchanged and, therefore, valid until the end of their respective validity duration. On the other hand, the update refers to the same weather event and should therefore achieve the same ActionID. In doing so, former DENMs are no longer transmitted due to the update. Consequently, the previous information, i. e., rectangular area, trace, generation time, and information quality, either get lost or are appended, e. g., into the *Alacarte Container*. However, the first choice facilitates a more precise distribution of the message, sharply adapted to the affected area. Therefore, this solution is assumed in context of this thesis.

Triggering Conditions

To detect weather events, use-case specific trigger conditions are evaluated. Based on consulted vehicle sensor data, respective quality levels are achieved and included in the DENM.

Within the advanced detection algorithm an information quality level is within a theoretical range of $0 - 10$. This information quality expresses the respective reliability of the current situation detection. Thereby, very low levels as well as very high levels might never be achieved, since the detection may always provide a minimal level of accuracy or will never gain enough evidences for a reliability value above a maximal level. However, this common set of information quality facilitates a high level of comparability between different weather types and an absolute appraisal within the possible range.

Within the advanced detection algorithm five different weather types are distinguished, i. e., *dense fog*, *heavy rain*, *heavy snowfall*, *aquaplaning*, and *slippery road*. In [PM07] a brief survey on how vehicle sensor data may be used to determine weather conditions is outlined. Expanding these basic suggestions, weather type specific trigger conditions and possible improvements of detection reliability as developed by the author of this thesis are detailed in the following.

To ensure dependable situation detection, the mandatory trigger conditions have to be fulfilled for at least the use-case specific minimal detection time t_{det}. Accordingly, to increase the information quality, the respective optional conditions have to be met during this time, in addition. However, if an optional condition is only temporarily fulfilled, the detection is not aborted. But if they are met a higher reliability level is reached.

Accordingly, a dense fog situation may be indicated by activating the low beam

Dense Fog			
Trigger Condition			**Quality**
	rear fog light	activated	2
and	low beam	activated	
and	vehicle speed	≤ 22 m/s	($\approx$ 80 km/h)
Optional Conditions			**Quality**
if	front fog light	activated	+1
if	relative humidity	$\geq 80\%$	+1
if	visibility	≤ 100 m	+2
if	vehicle speed	≤ 16 m/s	+1 ($\approx$ 60 km/h)
if	vehicle speed	≤ 11 m/s	+1 ($\approx$ 40 km/h)

TABLE 3.9.: Summary of mandatory trigger conditions to detect dense fog situations and optional conditions to increase detection reliability.

headlights and the rear fog light. In addition, high vehicle speeds may prove an unintended activation of these exterior lights. Hence, a maximum vehicle speed is assumed.

Moreover, vehicle speeds significantly below this level may increase the detection reliability. Additionally, by regarding the activation of the front fog light the detection quality might be further improved. Since fog appears as a result if the air's ability to absorb water is reached, e. g., due to decreasing temperatures, high relative air humidity proves presence of the dense fog at a high reliability. Moreover, by consulting on-vehicle cameras or lidar the meteorological visibility may be determined to confirm detection of fog situations [HLA05] [HTA07][32].

In Table 3.9 these trigger and optional conditions are summarized and associated quality level or improvement of detection reliability is proposed. Thereby, a minimal detection time t_{det} of 20 s is appropriate.

Participation events, i. e., heavy rain or heavy snowfall, are mainly indicated by the activation of the front shield wiper at a high level[33]. Additionally, to ensure a minimal detection reliability, low beam headlights have to be activated and the vehicle's speed has to be below a maximal detection speed. To distinct rainfall

---

[32] Measuring the meteorological visibility may additionally be detected, e. g., by a *Present Weather Detector*. Since such devices are unlikely available within vehicles this possibility is not further regarded in context of vehicle-based detection. However, infrastructure-based detection may include suchlike advanced detectors to determine meteorological visibility as outlined in Section 3.3.

[33] Respective front shield wiper speeds and levels within automotive context are specified for the European Union in [EC10].

Heavy Rain			
Trigger Condition			**Quality**
	wiper level	maximum	4
and	low beam	activated	
and	outside temperature	$\geq 2\,°C$	
and	vehicle speed	$\leq 28\,m/s$	($\approx 100\,km/h$)
Optional Conditions			**Quality**
if	rain sensor	$\geq 15\,mm/h$	+1
if	rain sensor	$\geq 25\,mm/h$	+1
if	vehicle speed	$\leq 22\,m/s$	+1 ($\approx 80\,km/h$)
if	vehicle speed	$\leq 16\,m/s$	+1 ($\approx 60\,km/h$)

TABLE 3.10.: Summary of mandatory trigger conditions to detect heavy rain situations and optional conditions to increase detection reliability.

from snowfall, the ambient temperature has to be verified. Snowfall appears in combination with low temperatures which implies a maximal detection temperature for that weather type. Accordingly, if the temperature is above the limits where snowflakes are formed, heavy rain is assumed.

To further enhance information quality, vehicle speeds significantly below the maximal detection speed might be regarded. A rain sensor, directly measuring the rainfall, may be regarded to further confirm the detection. Thereby, the level of snowfall is measured indirectly by volume of water from melted snowflakes on the front wind shield. An on-vehicle *lidar* may be consulted as a rain sensor, too [LEKK09]. Thus, rain- or snowfall is measured by calculating the *extinction coefficient* of radiated Laser waves.[34]

In Table 3.10 the trigger and optional conditions to detect heavy rain are summarized. Accordingly, Table 3.11 summarizes the trigger and optional conditions to detect snowfall. In both cases, a minimal detection time t_{det} of 20 s might be consulted, respectively.

Slippery road situations refer to areas with low traction due to ice or compressed snow on the roads. Such a condition might be detected during three driving maneuvers, i. e., on accelerating, on braking, and during driving. Thereby, the outside temperature has to be about the freezing point in all cases. Furthermore,

---

[34] In addition, a *disdrometer* may provide high reliable evidence on the presence of heavy rain or snowfall and, moreover, may trustworthy distinguish rain from snow. However, since the presence of disdrometers within vehicles is highly unlikely, it is not further regarded.

Heavy Snowfall			
Trigger Condition			**Quality**
	wiper level	maximum	4
and	low beam	activated	
and	outside temperature	$< 2\,°C$	
and	vehicle speed	$\leq 22\,m/s$	($\approx 80\,km/h$)
Optional Conditions			**Quality**
if	rain sensor	$\geq 10\,mm/h$	+1
if	rain sensor	$\geq 20\,mm/h$	+1
if	vehicle speed	$\leq 16\,m/s$	+1 ($\approx 60\,km/h$)
if	vehicle speed	$\leq 11\,m/s$	+1 ($\approx 40\,km/h$)

TABLE 3.11.: Summary of mandatory trigger conditions to detect heavy snowfall situations and optional conditions to increase detection reliability.

driving at high speeds indicates good road conditions and, therefore, prevents the detection.

An approach to partially recognize slippery roads is outlined in [HHSV07]. Based upon this, enhanced triggering conditions for all three driving maneuvers are adopted for the Weather Hazard Warning application.

If the throttle is pushed only slightly but a TCS is active anyway, a situation is indicated during this acceleration maneuver. In addition the vehicle acceleration may be regarded to ensure the detection. The detection may be further confirmed by temperatures below the freezing point and if the vehicle is already moving at higher speeds. Resulting trigger conditions for acceleration maneuvers including the optional conditions to increase the information quality are summarized in Table 3.12.

Analogously, if the braking pedal is pushed only slightly but the ABS is active anyway, a situation is indicated during this deceleration maneuver. As at acceleration, temperatures and vehicle's deceleration may be regarded to further confirm the detection. Respective trigger conditions for deceleration maneuvers are summarized in Table 3.13.

Finally, if the vehicle is neither accelerating nor decelerating, a slippery road situation may be indicated by an active ESP. However, hard driving maneuvers, e. g., driving through sharp curves at high speeds, generally force an ESP to keep the vehicle on track. Hence, such dynamic circumstances has to be excluded, e. g., by regarding the *side friction demand factor* f_D in relation with the current curve's

Slippery Road (at acceleration)			
Trigger Condition			**Quality**
	TCS	active	2
and	throttle	$\leq 30\%$	
and	outside temperature	$\leq 2\,°C$	
and	vehicle speed	$\leq 8\,\text{m/s}$	($\approx 30\,\text{km/h}$)
Optional Conditions			**Quality**
if	outside temperature	$\leq 0\,°C$	+1
if	vehicle speed	$\geq 2\,\text{m/s}$	+1 ($\approx 7\,\text{km/h}$)
if	vehicle acceleration	$\leq 1.5\,\text{m/s}^2$	+1
if	vehicle acceleration	$\leq 1.0\,\text{m/s}^2$	+1
if	vehicle acceleration	$\leq 0.7\,\text{m/s}^2$	+1

TABLE 3.12.: Summary of mandatory trigger conditions to detect slippery road situations at acceleration and optional conditions to increase detection reliability.

Slippery Road (at deceleration)			
Trigger Condition			**Quality**
	ABS	active	2
and	braking pedal	$\leq 40\%$	
and	outside temperature	$\leq 2\,°C$	
and	vehicle speed	$\leq 14\,\text{m/s}$	($\approx 50\,\text{km/h}$)
Optional Conditions			**Quality**
if	outside temperature	$\leq 0\,°C$	+1
if	vehicle deceleration	$\leq 3.0\,\text{m/s}^2$	+1
if	vehicle deceleration	$\leq 2.0\,\text{m/s}^2$	+1
if	vehicle deceleration	$\leq 1.2\,\text{m/s}^2$	+1

TABLE 3.13.: Summary of mandatory trigger conditions to detect slippery road situations at deceleration and optional conditions to increase detection reliability.

Slippery Road (during driving)			
Trigger Condition			**Quality**
	ESP	active	2
and	friction factor f_D	≤ 0.15	
and	throttle	$\leq 15\%$	
and	braking pedal	$\leq 20\%$	
and	outside temperature	$\leq 2\,°\mathrm{C}$	
and	vehicle speed	$\leq 14\,\mathrm{m/s}$	($\approx 50\,\mathrm{km/h}$)
Optional Conditions			**Quality**
if	outside temperature	$\leq 0\,°\mathrm{C}$	+1
if	friction factor f_D	≤ 0.10	+1
if	friction factor f_D	≤ 0.05	+1

TABLE 3.14.: Summary of mandatory trigger conditions to detect slippery road situations during driving and optional conditions to increase detection reliability.

curvature[35].

Hence, regarding the vehicles speed v and the Earth's standard acceleration g with

$$g = 9.80665\,\mathrm{m/s^2}\quad, \tag{3.9}$$

the friction factor is achieved by

$$f_D = \frac{v^2 \cdot \kappa}{g}\quad. \tag{3.10}$$

Thus, if the activation of the ESP is not explained by the calculated side friction factor f_D a slippery road is indicated[36].

In addition, to increase detection reliability, lower temperatures as well as lower side friction factors may be consulted. Resulting trigger conditions including the optional conditions are summarized in Table 3.14. Since no driver behavior, which might be done unintentionally, is included in these conditions, a comparable short detection time t_{det} of 1 s might be consulted in all three cases, respectively.

In contrast, aquaplaning situations refer to areas with low traction due to water on the roads. Such a condition might be detected within a braking or an appropriate

---

[35] A curve's curvature κ is the reciprocal of its radius r: $\kappa = r^{-1}$

[36] For intended speeds below 50 km/h curves are designed to facilitate side friction demand factors of $f_D \geq 0.15$, as summarized in [BON99]. This factor decreases with increasing curve radius [EBS04].

Aquaplaning (at deceleration)			
Trigger Condition		**Quality**	
	ESP or ABS	active	2
and	braking pedal	$\leq 40\%$	
and	outside temperature	$> 2\,°C$	
and	vehicle speed	≤ 25 m/s	(≈ 90 km/h)
Optional Conditions		**Quality**	
if	vehicle speed	≤ 19 m/s	+1 (≈ 70 km/h)
if	wiper level	medium	+1
if	rain sensor	≥ 10 mm/h	+1
if	low beam	activated	+1

TABLE 3.15.: Summary of mandatory trigger conditions to detect aquaplaning situations during braking maneuvers and optional conditions to increase detection reliability.

driving vehicle. Thereby, the outside temperature has to be significantly above the freezing point to enable liquid water instead of ice or snow on the road's surface.

Additionally, in both cases, an aquaplaning situation is indicated by activation of the ESP although the vehicle is driving adequate, i. e., not hard braking or driving at high speeds[37]. Moreover, lower speeds confirm the detection. In addition, front shield wipers, rain sensors, active low beam headlights, and low driving speeds may indicate rainy circumstances, which increases the probability of the presence of huge amounts of water on the road. According trigger and optional conditions are summarized in Table 3.15 for slightly braking vehicles and in Table 3.16 for driving vehicles. Since no driver behavior, which might be done unintentionally, is included in the trigger conditions, a comparable short minimal detection time t_{det} of 1 s might be consulted in both cases, respectively.

By consulting optical sensors, i. e., polarized light or laser, detection of road condition by means of ice and snow is possible in stationary cases within measurement stations [CRE12] [CKJ12] and might be possible within near future in mobile cases such as vehicles [IVSS10] [LHLS12]. Such sensors might be supplementary exploited in addition, once they are available.

The Weather Hazard Warning application as proposed in this thesis is intended to support situations regarding extreme weather conditions due to strong cross winds,

---

[37] Aquaplaning typically occurs during driving speed of 75–85 km/h, as, e. g., explained in [ADAC12].

Aquaplaning (during driving)			
Trigger Condition		**Quality**	
	ESP	active	2
and	friction factor f_D	≤ 0.15	
and	braking pedal	$\leq 20\%$	
and	outside temperature	$> 2\,°C$	
and	vehicle speed	$\leq 25\,m/s$	($\approx 90\,km/h$)
Optional Conditions		**Quality**	
if	vehicle speed	$\leq 19\,m/s$	+1 ($\approx 70\,km/h$)
if	friction factor f_D	≤ 0.10	+1
if	wiper level	medium	+1
if	rain sensor	$\geq 10\,mm/h$	+1
if	low beam	activated	+1

TABLE 3.16.: Summary of mandatory trigger conditions to detect aquaplaning situations during driving and optional conditions to increase detection reliability.

additionally. Unfortunately, there are no sufficient capabilities available in vehicles to reliable detect strong cross winds. Consequently, DENMs reporting strong wind situations are always generated by infrastructure-based detection within CIS, as further detailed in Section 3.3.

However, the advanced situation detection provides extensive precautions to avoid a positive detection if no event is present, i. e., *false-positive*, or no detection if the vehicle in fact passes an event, i. e., *false-negative*. To prove these techniques, they were implemented according detection algorithms within the Weather Hazard Warning applications of the field operational tests sim[TD] and DRIVE C2X, as further detailed in Section 6.3 and Section 6.4, respectively.

3.2.3. Determining a Surrounding Rectangle

Generally, a dangerous situation, e. g., a critical weather event, is spread over a large geographical area. Whenever a C2X application detects such a situation, the affected area has to be determined.

A strait forward way to do so is to record the driven way by means of trace-points while passing through the event. After the whole event is passed or periodically after certain time periods, a rectangle may be calculated that includes all of these

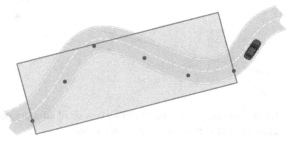

cf. [JH13]

FIGURE 3.3.: A rectangle surrounding a set of recorded positions, i. e., a recorded trace.

trace-points. In Figure 3.3 such a use-case is depicted.

This determined rectangle might be included as relevance area in a corresponding C2X messages for notifying nearby ITS stations. Thereby, an appropriate destination area is given, e. g., by just enlarging the rectangle by a suitable distribution offset.

An approach to determine such a rectangle was previously proposed in [JH13] by the author of this thesis. Hence, in this Section detailed instructions are proposed, how a rectangle may be calculated. In addition, the algorithm is implemented and evaluate within the Weather Hazard Warning applications in context of research projects as further detailed in Section 6.3 and Section 6.4.

To calculate a rectangle including a given set of arbitrary points[38], the following four steps have to be performed.

1. Since the sets of points may follow the course of a street, it is most likely that they are ordered in a line. Hence, the orientation of that chain of points has to be determined.

2. Since the line will not be perfectly strait, the width of the surrounding rectangle has to be calculated.

3. Afterwards, the length of the rectangle has to be determined.

4. Finally, out of these data, the parameters M, a, b, and θ as used in a DENM are calculated.

The main orientation of the given cloud of points is obtained by calculating a *regression line* through all points. Therefore, the method of *least squares* is applied.

---

[38] Even though the algorithm is motivated by the need to surround a trace, it is applicable for any arbitrary set of points.

cf. [JH13]

FIGURE 3.4.: Regression line g in the middle of a given set of arbitrary points, e. g.,
one or multiple recorded trace along a road.

Hence, for a set of n points $P_i = (x_i, y_i)$ the regression line g in a Cartesian plane is
defined by the linear equation in Slope-intercept form

$$g : y = m \cdot x + \gamma \quad . \tag{3.11}$$

Thereby, if $\bar{x}$ and $\bar{y}$ are the means of the x- and y-coordinates, respectively, the value
of the gradient m can be obtained by applying

$$m = \frac{\sum_i^n (x_i y_i) - n \cdot \bar{x} \bar{y}}{\sum_i^n (x_i^2) - n \cdot \bar{x}^2} \quad . \tag{3.12}$$

The value of the y-intercept γ is calculated subsequently by solving

$$\gamma = \bar{y} - m\bar{x} \quad . \tag{3.13}$$

In doing so, the main orientation of the arbitrary set of points is according to the
direction of this regression line. Additionally, the line is located in the middle of
the set, as depicted in Figure 3.4.

In contrast to the calculation of the regression line, which is done in affine coordinates, the width and the length of the rectangle is determined by transferring the
task into the vector space $\mathbb{R}^2$. Therefore, the regression line g should be depicted in
normal-vector-form

$$0 = \vec{P} \cdot \vec{n} - c \quad . \tag{3.14}$$

While using the numerator m_n and denominator m_d of the gradient m, the vector $\vec{v}$,
pointing into the direction of the line, is given by

$$\vec{v} = \begin{pmatrix} -m_d \\ m_n \end{pmatrix} = \begin{pmatrix} -(\sum_i^n (x_i^2) - n \cdot \bar{x}^2) \\ \sum_i^n (x_i y_i) - n \cdot \bar{x} \bar{y} \end{pmatrix} \tag{3.15}$$

and, thus, the normal vector $\vec{n}$ of the regression line g by

$$\vec{n} = \frac{\vec{v}}{|\vec{v}|} \quad . \tag{3.16}$$

Subsequently, the constant c is calculated by applying the equation

$$c = \begin{pmatrix} 0 \\ \gamma \end{pmatrix} \cdot \vec{n} \quad . \tag{3.17}$$

Hence, the distance d from an arbitrary point P, with position vector $\vec{P}$, to line g can be calculated with

$$d = \left| \vec{P} \cdot \vec{n} - c \right| \quad . \tag{3.18}$$

Thus, the width of the rectangle is achieved by calculating all distances d_i from any point P_i in the given set of points to the regression line g and multiplying the largest distance d_{max} by two.

$$d_{max} = \max \left(\{ d_i \mid \forall\, P_i \} \right) \tag{3.19}$$

$$width = 2 \cdot d_{max} \tag{3.20}$$

In contrast, to achieve the length of the rectangle, for every point P_i the corresponding foot-point F_i given by

$$\vec{F_i} = \vec{P_i} - d_i \cdot \vec{n} \tag{3.21}$$

has to be calculated in a first step. Within this set of foot-points, there is one point F_{max} with maximal and one point F_{min} with minimal x-component, respectively. These two points again define a vector $\vec{u}$ with

$$\vec{u} = \vec{F}_{max} - \vec{F}_{min} \quad . \tag{3.22}$$

Consequently, the length of the rectangle is given by the length of the vector $\vec{u}$

$$length = |\vec{u}| \quad . \tag{3.23}$$

Figure 3.5 summarizes and illustrates these steps to achieve the width and the length of a rectangle with foot-points F_{min} and F_{max}.

Finally, the four parameters to describe a rectangle within a DENM are calculated based on the achieved vectors, points, length, and width by just applying

$$M = F_{min} + \tfrac{1}{2} \cdot \vec{u} \tag{3.24}$$

$$a = \tfrac{1}{2} \cdot length \tag{3.25}$$

$$b = \tfrac{1}{2} \cdot width \tag{3.26}$$

$$\theta = \left(\tfrac{\pi}{2} - \arctan2(u_2, u_1) \right) \cdot \tfrac{180}{\pi} \quad , \tag{3.27}$$

whereas it has to be ensured that $0° \leq \theta < 360°$ holds, by adding $360°$ to θ if θ is negative.

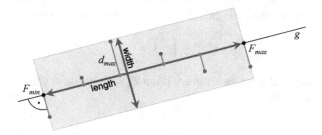

cf. [JH13]

FIGURE 3.5.: Calculation of the width and the length of a rectangle based on the distance d_{max} and the distance between the foot-points F_{min} and F_{max}.

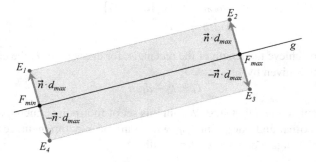

FIGURE 3.6.: Calculation of the four edges E_1 to E_4 of a rectangle by adding and subtracting $\vec{n} \cdot d_{max}$ to/from F_{min} and F_{max}, respectively.

However, due to the assumed large geographical expansion of hazardous situations, extremely small rectangles should be avoided. Consequently, within the Weather Hazard Warning application a minimum length and minimum width of, e. g., 100 m is ensured for rectangles.

Finally, if needed, the four edges E_1 to E_4 of the rectangle can be calculated by adding and subtracting $\vec{n} \cdot d_{max}$ to/from F_{min} and F_{max}, respectively, as depicted in Figure 3.6.

3.3. Infrastructure-Based Weather Detection

To supplement detection of the current weather situation carried out within vehicles, additional information is gathered and evaluated within infrastructure stations such as *Roadside ITS Stations* (RIS) or *Central ITS Stations* (CIS). For Car-to-X communication not only one central, but multiple such local weather centers are required. This infrastructure-based detection is not within the main focus of this thesis. However, to facilitate an overview on the entire spectrum of the Weather Hazard Warning application within C2X communication, a short outline is given in this section.

Primarily data sources for infrastructure-based weather situation detection are weather measurements stations located nearby roads. Currently, a well-developed infrastructure of advanced weather measurement stations is already installed and maintained by local or national authorities or by private meteorological service providers. By connecting these already existing measurement station to ITS stations, reliable warning messages might be calculated directly by applications within RISs and CISs and provided to vehicles nearby.[39]

Equipped with a huge set of various highly sophisticated sensors, weather measurement stations deliver highly precise measurements such as, e. g.,

- air temperature,
- dew-point temperature,
- street surface temperature,
- in-earth temperature,
- precipitation intensity and type,
- street surface state and layer thickness,
- atmospheric pressure,
- relative humidity,
- wind speed and direction,
- sight distance, and
- remaining salt ratio on the street.

Based on this various set of different sensor values, infrastructure stations may detect weather situations such as *dense fog*, *heavy rain*, *heavy snowfall*, *aquaplaning*, *slippery road*, and *strong cross winds*.

Therefore, the advanced detection algorithm as detailed in Section 3.2.2 and illustrated in Figure 3.2 might be consulted for an event detection. Since infrastructure stations do not move, no trace has to be maintained, obviously. Instead, a static pre-defined trace and rectangle can be applied. Other DENM parameter may be set according to Section 3.2.2.

---

[39] In addition, further conventional information sources as rain radar and satellite images may be provided and consulted within the detection [PM07]. However, this is far beyond the context of this thesis and is not further regarded.

Weather Type		Trigger Condition	
Dense Fog		sight distance	$\leq 100\,$m
	and	relative humidity	$\geq 80\,\%$
Heavy Rain		precipitation intensity	$\geq 20\,$mm/h
	and	precipitation type	rain
	or	air temperature	$\geq 2\,°\mathrm{C}$
Heavy Snowfall		precipitation intensity	$\geq 20\,$mm/h
	and	precipitation type	snow
	or	air temperature	$< 2\,°\mathrm{C}$
Aquaplaning		street surface state	wet
	and	street surface layer thickness	$\geq 7\,$mm
Slippery Road		street surface state	ice
	or	street surface state	snow
	and	street surface layer thickness	$\geq 2\,$mm
	and	street surface temperature	$\leq 0\,°\mathrm{C}$
Strong Cross Winds		wind speed	$\geq 14\,$m/s
	and	wind direction	orthogonal

TABLE 3.17.: Summary of trigger conditions to detect adverse weather situations on infrastructure stations, i. e., Roadside ITS Stations or Central ITS Stations.

For infrastructure-based detection, trigger conditions have to be adopted and possible simplified trigger conditions are proposed in Table 3.17. In addition, the update, detection, and repetition time may be adjusted to

$$t_{upd} = 10\,s \quad , \tag{3.28}$$

$$t_{det} = 30\,s \quad , \text{and} \tag{3.29}$$

$$t_{rep} = 60\,s \quad , \tag{3.30}$$

respectively. Thereby, detection of strong cross winds is intended, which is not supported by vehicle-based detection as outlined in Section 3.2.2. However, detection of winds needs extensive adaption of the algorithm and its parameters, since multiple and heavy but temporary limited strong cross winds have to be detected.

Weather station measurements are highly reliable but are valid for the local area only. Hence, by extrapolating the measurements along the road, a larger affected area may be determined by the cost of reliability. Consequently, each station may generate multiple messages around each measurement station. Thereby, reliability

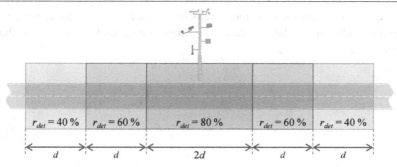

FIGURE 3.7.: With increasing distance to the weather measurement station, the detection reliability r_{det} of an infrastructure-based weather situation detection decreases.

is decreasing with increasing distance.

Accordingly, with the detailed algorithm and trigger conditions a reliability of approximately 80 % of the maximum value is proposed[40]. With increasing distance, in use-case specific step ranges d, to the measurement station, this reliability level decreases in, e. g., three steps, as depicted in Figure 3.7.

Furthermore, more advanced detection algorithms and forecasts are imaginable. Hence, measurements values may be observed and regarded more extensive, e. g., changes in air humidity facilitate prediction of rainy conditions [KRCS08], estimating[41] dew-point τ to predict dense fog [SON90] by regarding air humidity H_R and current temperature T with

$$\tau \approx T - \frac{100 - H_R}{5} \tag{3.31}$$

is possible [OBL04] or salt on the road may be regarded to calculate the possibility of frozen roads [TUR97]. In doing so, a maximum reliability level may be reached most likely.

In addition, also vehicles may be used to serve as mobiles sensors. Therefore, whether profiles gathered in vehicles as detailed in Section 3.1.3 are wrapped in PVDMs as proposed in Section 2.3.3 and transmitted to infrastructure stations. These probe vehicle data heavily extent information achieved from weather measurement stations. Hence, a more accurate, reliable, and complete overall weather view for the entire region is achieved. This way, detection on both sides, vehicles and infrastructure, may confirm each other and, hence, reliability is increased.

---

[40] According to Section 3.2.2, this implies a quality level of 8.
[41] This approximation is accurate for relative humidity above 50 % within about ±1 °C.

Moreover, since weather station measurements are only valid for local area, consulting data gathered over large distances facilitate more complex and reliable short time forecasts.

4. Situation Maintenance

As detailed, within Car-to-X communication each vehicle may detect and report dangerous situations. In Section A.1, some possible attacks on C2X use-cases are detailed. Consequently, multiple approaches to verify the trustworthiness of a message have to be applied upon message reception.

Accordingly, countermeasures regarding the lower layers of the C2X communication stack are detailed in Section A.2 and Section A.3. However, within the following of this chapter an in-depth analysis of messages content is proposed.

Additionally, since several C2X applications, like the Weather Hazard Warning application, detect and report dangerous situations in a successive manner, each DENM concerns just a small part of the whole event. Consequently, a receiving ITS station has to be aware of this relation between multiple received DENMs in order to determine the entire expansion of an overall event. Additionally, since multiple ITS stations may detect parts of the same event independently from each other, neighbored ITS stations will receive multiple messages regarding either the same event or the same part of the event. Such a situation is depicted in Figure 4.1, whereas two vehicles successively report partially overlapping situations within the same area.

Within an application, this relation between the multitude of single situations has to be relieved in order to determine the distribution of the respective real world event. This way, a driver notification can be triggered according to the overall event as further detailed in Chapter 5. Hence, the underlying application realization, based on multiple, independent, and successive situation detections, is hidden to the driver and only one notification per event is displayed.

Consequently, the aggregation of situations to an overall event is detailed afterwards. Thereby, concepts to determine the overall event area and the common event reliability as well as approaches to maintain known situations and events are presented.

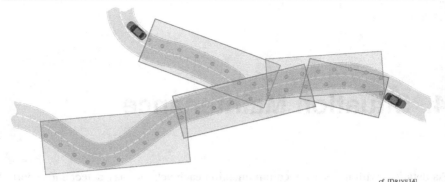

cf. [DRIVE14]

FIGURE 4.1.: Multiple partially overlapping situations of one successively reported
large event, independently detected by two different ITS stations.
For each situation measurement point and surrounding rectangles
are depicted.

4.1. Mobility Data Verification

Almost every traffic safety and traffic efficiency related application in Car-2-X
communication systems depends on correct mobility information of adjacent ITS
stations. As detailed in context of the analysis of possible attacks to C2X applications at the beginning of this chapter, faked messages injected by a stationary
attacker located on the roadside is identified as the most likely and threatening
attack. As discussed in the attack on the *Green Light Optimal Speed Advisory* application, a moving attacker is identified as rather threatening, because its motivation
may be higher compared to attackers that are not part of the road traffic. Therefore,
sender authentication and message integrity are cryptographically secured within
C2X communication [IEEE06] as detailed in Section A.3.

Additionally, in order to enhance trustworthiness of mobility information received
from neighbored ITS stations, IEEE 1609.2 proposes a *plausibility validation* of
message content. Thereby, application specific information should be verified by the
respective C2X application as outlined for the Weather Hazard Warning application
in Section 4.2. In contrast, vehicle mobility information is not related to specific
but to various applications and system components. Especially, the necessity of
mobility data plausibility verification in C2X communication is also investigated
in [LSK06]. Due to the importance of correct ITS station mobility information,
detection of misbehavior in C2X communication is considered in several different
approaches.

One concept, e. g., proposes to exploit omnidirectional radar sensors to verify position information from ITS station in the direct neighborhood [YCWO07]. Due to the strongly limited observation range of radar sensors, a routing topology that allows the usage of radar information cooperatively in the neighborhood is proposed. However, in doing so new capabilities to inject faked data are introduced.

In [RLY+09] a relative location verification protocol is proposed. Thereby, directional antennas are exploited to at least distinguish between vehicles in front and behind.

Another position verification approach is based on communication modules that feature *Received Signal Strength Indicator* (RSSI) [XYG06] [LB09]. This allows to calculate the sender's distance based on a radio model. Nevertheless, this technique for position estimation heavily lacks of accuracy. Hence, it is not used for direct distance measurements but is proposed, to apply an analysis of signal strength distribution indicating where the signal origins from.

All these concepts for position information verification suffer from inherent inaccuracies and essential high technical requirements.

In contrast, within the *Vehicle Behavior Analysis and Evaluation Scheme* (VEBAS) several basic checks to identify faked position claims are proposed [SLS+08]. Thereby, messages containing mobility data which exceeds physical boundaries, e. g., speeds above 300 km/h, are excluded. However, this approach does not facilitate sufficient trustworthiness since attackers may easily meet these boundaries even in faked messages.

Consequently, the author of this thesis co-develops a component for *Mobility Data Verification* relying on information included in received messages and impeding generation of faked mobility data by additional movement analysis. This component is firstly proposed in [SJB+10], described in [JBSH12] and detailed in this section. Thereby, the author of this thesis has designed the verification flow and developed and implemented the core part of the component, i. e., the vehicle trackers and their management.

In order to evaluate the presented component, it is introduced as part of the security solution of the German research project sim^TD [SIMTD10a] as detailed in Section 6.2. Additionally, enhanced checks based on vehicles local sensors and an improved verification flow is propose and discussed, subsequently.

The *Mobility Data Verification* thereby evaluates data that are common to every C2X message. Hence, these checks should be composed in a common component in order to avoid multiple validations. Additionally, messages have to be decoded to extract and evaluate contained information. Hence, it has to be integrated into the C2X communication system after these steps are done.

The system component *Local Dynamic Map* (LDM), located on the facility layer of the C2X communication stack, provides all received messages to the C2X application. Thereby, the LDM evaluates message content on a low level to facilitate functionality to filter messages according to application requirements. Accordingly, to integrate the component into an overall C2X system architecture, it may be attached onto the *Local Dynamic Map*. This way, passing untrustworthy messages to applications is prevented and filter capabilities are enhanced.

Within *Mobility Data Verification* evaluation of mobility data is performed upon every received C2X message in a serial order. Thereby, unrealistic data are excluded by applying multiple *threshold checks*, in a first step. Hence, mobility data, which exceed physical boundaries, are filtered.

Subsequently, a Kalman filter-based approach to estimate a vehicle's future movements, which serve as a basis for mobility verification is advocated. Based on the sending *vehicle's ID* messages are mapped to according trackers.

Taking into account privacy considerations, vehicles may frequently change their identifiers, as detailed in Section 2.4. These *Pseudonym Changes* complicate mobility verification significantly. Nevertheless, the introduced approach provides reliable results even in presence of such loss of all identifiers.

Finally, sensor integration is proposed. Thereby, additional vehicle sensors, such as radar used in ACC, are consulted to evaluate vehicles in the near environment more reliable.

After processing, each message will be classified as *Approved*, *Neutral*, or *Erroneous*.

The Mobility Data Verification consults a Kalman filter-based vehicle tracker to verify the mobility data within received messages. In the following a brief introduction into Kalman filter theory is given. Subsequently, the overall verification process is detailed.

4.1.1. The Kalman Filter

To estimate the behavior of a linear system, a Kalman filter [KÁL60] may be advocated [WEN07]. Especially for object tracking, a Kalman filter represents an efficient and well-known solution [BP99]. Thereby, the filter repeats two successive phases for every time step k, the prediction and the correction.

Within the first phase, the *Prediction*, a prediction $\hat{x}_k$ of the *system state* for this time step is calculated by multiplying the last predicted state $\hat{x}_{k-1}^+$ with the *state transition matrix* F_k. The state transition matrix, thereby, is the mathematical representation of the underlying system model, e. g., a mobility model of a vehicle.

$$\hat{x}_k = F_k \cdot \hat{x}_{k-1}^+ \tag{4.1}$$

However, to get a more accurate prediction, additional *control values* u_k, may be added via a *control matrix* B_k to the system state before the state transition matrix is applied. This way, inaccuracies of the system model can be compensated or further information, which is not regarded in the system model, can be inserted, respectively.

$$\hat{x}_k = F_k \cdot (\hat{x}_{k-1}^+ + B_k \cdot u_k) \tag{4.2}$$

In addition, an estimated *prediction error* P_k for the system state prediction is calculated based on the system model, the last time steps prediction error, and the *system fault matrix* Q_k. This system fault matrix represents errors that are inherent in the used system model and therefore affect the accuracy of the prediction.

$$P_k = F_k \cdot P_{k-1}^+ \cdot F_k^T + Q_k \tag{4.3}$$

In order to achieve a more accurate system state, within the following second phase, the *Correction*, the predicted state is adjusted by means of measurement values. Hence, the *difference* Δy_k between *measured values* $\tilde{y}_k$ and *predicted measurements* $\hat{y}_k$ is calculated. Predicted measurement values are achieved based on the current system state, by applying a *measurement matrix* H_k to the system state.

$$\Delta y_k = \tilde{y}_k - H_k \cdot \hat{x}_k \tag{4.4}$$

Additionally, the *Kalman gain* K_k is calculated based on the prediction error and the *measurement variances* R_k that are inherent to the used measurement technique. Thereby, the relation between system state and measurement values, as represented by the measurement matrix, is taken into account.

$$K_k = P_k \cdot H_k^T \cdot (H_k \cdot P_k \cdot H_k^T + R_k)^{-1} \tag{4.5}$$

Hence, the *corrected system state* $\hat{x}_k^+$ is established by weighting the difference Δy_k with the Kalman gain and adding it to the system state.

$$\hat{x}_k^+ = \hat{x}_k + K_k \cdot \Delta y_k \tag{4.6}$$

Finally, the prediction error is corrected, too, in order to achieve a more accurate prediction error P_k^+.

$$P_k^+ = P_k - K_k \cdot H_k \cdot P_k \tag{4.7}$$

Corrected system state and prediction error are used in the succeeding prediction phase at time step $k+1$.

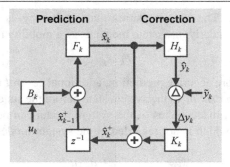

cf. [JBSH12]

FIGURE 4.2.: Structure of the Kalman filter, with respect to the prediction and the
correction phase as well as used matrices and operations.

Validation Class	Description
Erroneous	The mobility data do not match the assumed vehicle dynamics.
Neutral	The component cannot make a reliable statement.
Approved	The mobility data are checked and accepted.

cf. [JBSH12]

TABLE 4.1.: Summary and description of the three validation classes provided by
the Mobility Data Verification component to rate trustworthiness of
C2X messages by means of vehicle dynamics.

The schematic in Figure 4.2 illustrates the two Kalman filter phases. Thereby, z^{-1}
denotes the time shift between time step $k-1$ and k or k and $k+1$, respectively.

4.1.2. Mobility Data Verification Flow

The *Mobility Data Verification* component evaluates vehicle mobility data contained
in the header of every C2X message. By verifying that received mobility data are
consistent with typical vehicle dynamics, each C2X message will be classified
as either *Approved*, *Neutral*, or *Erroneous*. These possible results can easily be
interpreted and used by applications and system components on the corresponding
vehicle. The three validation classes are summarized and detailed in Table 4.1

In context of the Mobility Data Verification, *vehicle mobility data* refer to the
vehicle's

- position (longitude, latitude),
- speed, and
- heading.

Thereby, the component performs evaluation upon every single received C2X message successively. Thereby, mobility data and sender ID are extracted from the message and processed.

In the first step of the verification flow multiple *threshold checks* are performed. Thereby, mobility data, which exceed basic physical boundaries for vehicle dynamics or radio communication, are filtered. As introduced in [SLS$^+$08], these basic checks include a verification of the vehicle's speed, since even in highway scenarios vehicles may only drive with a maximum velocity. Furthermore, a message which indicates a sender's position outside of the host vehicle's communication range[42] is regarded as untrustworthy. Another threshold check monitors the repetition frequency of messages, especially CAMs, which have to meet the maximum repetition frequency stated in Section 2.3.1[43]. Additionally, the message's time stamp is checked to detect messages that are either expired or dated to a future point in time[44]. Besides filtering faulty measurements as well as very simple designed faked messages, the stated threshold verification prevents inconsistent data that may corrupt the subsequent mobility prediction.

The threshold checks are summarized in Table 4.2. As those checks are very basic and with huge margins, each of them has to be passed successfully. Hence, if only one of these checks fails, the message is instantly marked as *Erroneous* and the verification process is aborted.

If all threshold checks are passed successfully the message must undergo an in-depth analysis of vehicle dynamics. Therefore, the author of this thesis chooses Kalman filter-based vehicle trackers to predict mobility data of adjacent vehicles. Such a filter represents an efficient solution for object tracking [BP99].

Furthermore, a Kalman filter-based vehicle tracker easily overcomes pseudonym changes [WMKP10]. Therefore, this approach is most appropriate for usage in C2X communication scenarios. In Section 6.2 the adaption of the filter for the purpose of vehicle tracking is detailed.

For each known vehicle within communication range an individual tracker is maintained within the component. Thus, newly received messages are evaluated with respect to previous messages from the corresponding vehicle, i. e., it is observed if all messages lie on a continuous trace in compliance with possible vehicle dynamics.

---

[42] In order to verify the communication range, own vehicles positioning information is maintained in the component.

[43] In doing so, a *Denial-of-Service* (DoS) attack by stressing the system with a huge amount of messages is impeded.

[44] Since some messages are forwarded and valid for a long time as detailed in Section 2.3.2 and Section 2.3.4, the time stamp added by the last sender is regarded.

Threshold Check	Limit	Description
vehicle speed	83 m/s	This check verifies, if the vehicle speed is within the limits of current vehicles maximum speeds. ($\approx$ 300 km/h)
reverse speed	−14 m/s	For reverse driving vehicles, the speed is negative and the limits are more restricted. ($\approx$ −50 km/h)
acceptance range	1000 m	The senders should not be positioned outside of the maximum communication range facilitated by C2X communication.
message frequency	10 Hz	Successive messages from one vehicle must not be transmitted with a frequency above the specified limit.
time stamp	500 ms	Sending time of received messages should already be passed but not older than usual transmission takes.

TABLE 4.2.: Summary, limits, and description of the threshold checks applied by the Mobility Data Verification component.

Mobility Data	Limit	
position	1.5 m	
speed	1.5 m/s	($\approx$ 5 km/h)
heading	4°	

TABLE 4.3.: Summary of the thresholds applied by the Mobility Data Verification component on comparing predicted and received mobility data.

Hence, after passing the threshold checks, the corresponding tracker is identified by the respective vehicle ID. Assuming that the component receives a message from an already known vehicle, a matching tracker is found and its prediction phase, as detailed in Section 4.1.1, is triggered. Thereby, since the filter's state $\hat{x}_{k-1}^{+}$ represents the last known vehicle mobility data, the elapsed time t_k between last received message and current message is regarded in the formulas. Accordingly, the new predicted state $\hat{x}_{k}^{+}$ represents the expected mobility data and, therefore, is compared with the received message's content.

Since the filter only provides an estimation of the vehicles mobility data, they will never match perfectly. However, if the derivation between expected and received data is below acceptable limits, the message is assessed as trustworthy and, consequently, evaluated as *Approved*.

An appropriate approach to define those limits consults static pre-defined thresholds as proposed in Table 4.3. However, more advanced concepts regard the previous calculated *Kalman gain K_{k-1}* provided by the Kalman filter. Since the Kalman gain gives evidence on the accuracy of the working filter, it describes how precise the provided prediction and how large possible errors are. Consequently, with low quality rating, stated limits for acceptance might be increased.

Finally, since the received mobility data are verified, the correction phase of the Kalman filter is applied as detailed in Section 4.1.1. Thereby, the filter's system state is improved by means of new measurement values, i. e., the received data.

In contrast, if the derivation in this last check is not within acceptable limits, the message is evaluated as *Erroneous*. Accordingly, changes applied to the Kalman filter are reversed to ensure a consistent state and the validation flow is finished.

If the threshold checks at the beginning of the verification flow are passed but the received vehicle ID is unknown, no vehicle tracker will be found. However, if an already known vehicle has performed a pseudonym change, as discussed in Section 2.4, a matching vehicle tracker still exists within the Mobility Data Verification component. Hence, the pseudonym change is made locally transparent within the host vehicle.

In order to find a candidate tracker which corresponds most likely to the sending vehicle, the prediction of all known trackers is preliminarily tested to fit the received mobility data. The most feasible tracker, i. e., the one whose prediction matches the received mobility data the best, is selected as candidate. If the vehicle movement fits the prediction of this tracker, i. e., the derivation is within the expected limits as stated in Table 4.3, a pseudonym change is considered to be resolved. Accordingly, the trackers vehicle ID is updated, the correction phase of the filter is executed, and the message is finally evaluated as *Approved*.

Additionally, if needed, a *local node ID* might be assigned to the message, which is mapping both vehicle IDs to a unique identifier. This way, the pseudonym change is made transparent to local applications and system components, such that they may continue their work unharmed. However, pseudonym changes are performed for reasons of driver's privacy. Hence, this mapping of two different pseudonyms has to be stored just temporarily and used locally within the host vehicle, only.

However, if the deviation between received data and the prediction of the most feasible tracker does not meet the set boundaries, no pseudonym change is considered, instead a new appearing vehicle is assumed. Since a pseudonym change is excluded, additional evaluations adequate to evaluate a new appearing vehicle have to be done to classify the trustworthiness of the message.

Generally, previous unknown vehicles firstly appear at the border of the host vehicles communication range. Regarding that vehicles move while sending messages and

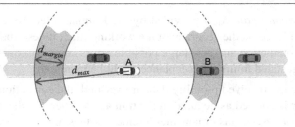

cf. [JBSH12]

FIGURE 4.3.: The *Appearance Margin Check* verifies a suddenly appearing vehicle B only if it is firstly noticed within a margin d_{margin} near the border of the vehicle's maximum communication range d_{max}.

that messages may get lost, especially on the outer end of the communication range, new appearing vehicles are noticed within a certain margin. Consequently, an *Appearance Margin Check* is applied. This rather light-weighted test accepts all new appearing vehicles inside the current maximum communication range d_{max}, but only if they are within a tolerance margin d_{margin} of, e. g.,

$$d_{margin} = 150\,\text{m} \quad . \tag{4.8}$$

Such a case is illustrated in Figure 4.3, where vehicle A accepts a first received message from vehicle B. In contrast, first messages from the other two depicted vehicles would fail the test.

Nevertheless, starting vehicles may validly appear near the host vehicle. Taking into account that these vehicles may drive at relatively low speeds, they are accepted by the Appearance Margin Check anyway, if their velocity is below a threshold v_{max} of, e. g.,

$$v_{max} = 1.5\,\text{m/s} \quad (\approx 5\,\text{km/h}) \quad . \tag{4.9}$$

Accordingly, if the message fulfills these tests for suddenly appearing vehicles, a valid new appearing vehicle is assumed. Hence, a new vehicle tracker, associated with the vehicle's ID, is instantiated. Since there exists no information about the previous path and driving dynamics of that vehicle, no reliable decision on the validity of the received message is possible. Hence, the validation flow is terminated with classifying the message as *Neutral*.

Nevertheless, a fail of the *Acceptance Margin Check* still does not necessarily indicate a faked message. In Figure 4.4 a common highway scenario is depicted. Thereby, vehicle B is starting to overtake the truck in its front. Due to shadowing effects, caused by the truck in the rear, previous C2X messages originated by B are blocked to the approaching vehicle A. In consequence, vehicle B will fail the

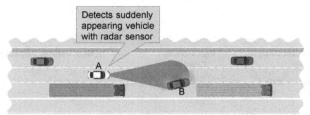

cf. [JBSH12]

FIGURE 4.4.: Within critical situations suddenly appearing vehicles, which fail the acceptance margin range check due to, e. g., shadowing, are verified by consulting local vehicle sensors, additionally.

Acceptance Margin Check, as vehicle A cannot distinguish this scenario from the one depicted in Figure 4.3.

Therefore complementary checks based on vehicle's local sensors, e. g., radar- or camera-based approaches, are proposed. With such sensors the distance and angle of vehicles driving ahead are constantly measured. Accordingly, detected objects may be matched to positions indicated by C2X messages. Thereby, the different coordinate planes as well as error variances of both systems are taken into account. By consulting this concept, a *Local Sensor Check* providing reliable information for at least vehicles in direct *Line-of-Sight* (LoS) is facilitated. In Section 6.2.2 a prototypical implementation of such a *vehicle sensor data fusion* is outlined. Additionally, to finalize the verification flow, a corresponding vehicle tracker is instantiated, such that a matching tracker will be found directly at the beginning of the upcoming verification.

For any vehicle firstly indicated in non-LoS but within the near communication range of the host vehicle, the first message has to be evaluated as *Erroneous*. Nevertheless, a new matching vehicle tracker is instantiated, such that future received messages might be successfully evaluated.

However, most of the safety related use-cases, like the *Forward Collision Warning*, operate on vehicles in direct LoS. Hence, these use-cases, are facilitated since valid messages are evaluated as *Approved*.

The entire mobility data verification flow as detailed in this section is depicted at a glance in Figure 4.5.

4.2. Weather Data Verification

In the previous section of this chapter the author of this thesis details a technique to enhance trustworthiness of mobility information within C2X messages. In contrast,

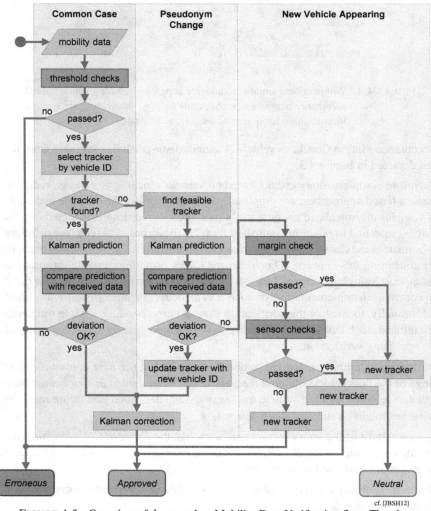

cf. [JBSH12]

FIGURE 4.5.: Overview of the complete Mobility Data Verification flow. Thereby,
           three partially successive cases, i. e., the common basic case, a
           pseudonym change, and a new appearing vehicle, are regarded.

Vehicle Data	Limit	Description
wiper speed	≤ 60 wipes/min	Frond shield wipers will not provide more than 1 wipe per second.
curvature	$\leq 0.1\,\mathrm{m}^{-1}$	Curves with a radius of less than $10\,\mathrm{m}$ are to narrow for road traffic.
acceleration	$\leq 5\,\mathrm{m/s}^2$	Vehicles will not accelerate with more than $\frac{1}{2}\,g$.
deceleration	$\leq 7.5\,\mathrm{m/s}^2$	Vehicles brake harder than they may accelerate.
vehicle speed	$\leq 83\,\mathrm{m/s}$	Even on highways, vehicle's speed is below a maximum limit. ($\approx 300\,\mathrm{km/h}$)
air temperature	$\leq 70\,^{\circ}\mathrm{C}$	Maximum realistic temperature on earth,
	$\geq -30\,^{\circ}\mathrm{C}$	and minimum realistic temperature on earth.

TABLE 4.4.: Possible thresholds for several selected vehicle data consulted within the situation detection of the Weather Hazard Warning application as detailed in Chapter 3.

application specific message content is exuded from previous evaluation and, hence, has to be verified by appropriate evaluations. As, e. g., outlined in [IEEE06] and Section 4.1, this should be verified best by the corresponding application.

Obviously, application specific data highly depend on the respective application as well as suitable verification approaches. Hence, a description of verification schemes integrated at the *Application* layer is done exemplary. Accordingly, in this section the author of this thesis introduces concepts to add message reliability directly integrated within the *Weather Hazard Warning* application.

Verification of application specific data may be provided from two complementary directions. It is not performed on incoming messages, only, but also on data provided from the host vehicle. Since vehicle internal data may be corrupted due to sensor malfunction or by an adversary as outlined in Section A.1.1, additional reliability is achieved by verifying data before faulty or faked messages are sent or even before detection algorithms are triggered.

According to the *Mobility Data Verification* as detailed in Section 4.1.2, weather related vehicle data may be firstly checked against physical and technical boundaries, too. Such possible boundaries for selected vehicle data regarded within the Weather Hazard Warning application as enumerated in Chapter 3 are listed in Table 4.4.

Nevertheless, such basic plausibility checks on host's vehicle data do not avoid false-positive detections, faked but plausible vehicle data or, even less, intentionally injected C2X messages. Hence, received message's content still has to be checked comprehensively for plausibility.

Situation Type	Local Vehicle Data	Quality
dense fog	temperature ≥ 10 °C	−1
	temperature ≥ 15 °C	−1
	temperature ≥ 20 °C	−1
	relative humidity ≤ 70 %	−2
heavy rain	low beam deactivated	−1
	relative humidity ≤ 50 %	−1
	temperature ≤ −1 °C	snowfall
heavy snowfall	low beam deactivated	−1
	temperature ≥ 5 °C	rain
slippery road	temperature ≥ 3 °C	−1
	temperature ≥ 7 °C	−2

TABLE 4.5.: Recommended cross-checks for several selected received weather
situations by means of local sensor data.

Thereby, a simple three-staged classification scheme as applied in context of the
Mobility Data Verification is not sufficient. Even more, since the Weather Hazard
Warning application handles DENMs containing a value for information quality,
this reliability value may be adjusted due to plausibility evaluation.

First of all, received situations may be cross-checked by local vehicle data. Thus,
the reliability of, e. g., a received slippery road or heavy snowfall situation may be
decreased, if the local temperature indicated rather high temperatures. It further
might be considered, e. g., that dense fog, heavy rain, or heavy snowfall events
might force activation of low beam lights already within near environment. Selected
suchlike cross-checks against local sensor data with respective recommend effect
on the information quality is detailed in Table 4.5. However, since the vehicle is
notified about a weather situation not jet reached, these plausibility cross-checks
will always indicate a tendency of trustworthiness, only.

Moreover, weather events appear, move, and disappear over time. In contrast, an
once transmitted DENM stays valid for its respective validity duration. This general
condition is, on the one hand, regarded within the message's validity duration.
However, in general a weather event does not disappear from one second to another
like the validity duration expires.

In order to additionally consider the weather event's changes over time more
precisely, the messages reliability is decreased over message's validity time. Such
an approach is deeply integrated within the application's situation management and,
hence, further detailed in Section 4.4.1

Additionally, the Weather Hazard Warning in C2X communication is intended to be a cooperative application. Hence, to increase reliability individual vehicle stations may confirm each other's detection.

Thereby, receiving stations generally do not trust single situation detections. If only one of multiple vehicles report a situation, it is most likely that this message is based on a false-positive detection, a sensor malfunction, or a faked message. Consequently, the Weather Hazard Warning demands multiple detections originated from multiple stations before the information is regarded as trustworthy. Moreover, by cross-checking messages by other messages, it could be assumed that the reliability in fact increases, if multiple individual ITS stations report the same event. Such approaches again are deeply integrated within the application's functionality of weather situation aggregation and, hence, are detailed in Section 4.4.2.

Supplementing, *Verify-on-Demand* concepts [KW11] may be adapted to application specific plausibility evaluations. By adjusting thresholds more strictly if, e. g., originating stations attract attention by sending multiple erroneous messages. Thereby, plausibility evaluation from lower levels may be considered as well as concepts of a per vehicle trust management as, e. g., proposed in [QUA11] or [SLS+08].

By applying such plausibility evaluations on application specific message content, plausibility verification even on application layer as claimed by current standards [IEEE06], is facilitated. Hence, the reliability of C2X applications may be significantly enhanced and, thus, the acceptance and trustworthiness of the entire C2X communication system.

4.3. Area Aggregation

Within an ITS station, the first step to be performed in order to aggregate two situations is to determine, if the two areas overlap. Within the Weather Hazard Warning application, situation areas are represented by rectangles. In order to determine overlapping rectangles, it has to be investigated, if their borders cross each other and if their edges are located inside the respective other rectangle. Hence, the presented algorithm relies on two subroutines, one verifying the borders and the other verifying the edges. These subroutines are detailed in the following, before the algorithm applied to determine the overlapping area is detailed.

4.3.1. Intersection Points of Bounded Lines

In this section an algorithm to determine if two bounded lines, as the borders of rectangles are, cross each other is detailed. The presented approach is outlined by the author of this thesis in [JH13] and implemented, tested, and verified within the *Road Weather Warning* application and the *Weather Warning* application in context of the research projects sim^{TD} [SIMTD] and DRIVE C2X [DRIVE], respectively.

In the $\mathbb{R}^2$ the vector equation defining an arbitrary line h is given by a point P and a vector $\vec{u}$ by

$$h : \vec{x} = \vec{P} + \lambda \vec{u} \quad . \tag{4.10}$$

The intersection point S of two lines h and h' is given by

$$\vec{S} = \begin{pmatrix} p_1 \\ p_2 \end{pmatrix} + \lambda_S \begin{pmatrix} u_1 \\ u_2 \end{pmatrix} \quad , \tag{4.11}$$

if there is a non-trivial solution for λ_S in

$$\lambda_S = \frac{p_1 u_2' - p_2 u_1' - p_1' u_2' + p_2' u_1'}{u_1' u_1 - u_2' u_2} \quad . \tag{4.12}$$

Subsequently, λ_S' is achieved by calculating

$$\lambda_S' = \frac{s_1 - p_1'}{u_1'} \quad . \tag{4.13}$$

However, such general lines have infinite length and their intersection points are located along the lines without restrictions. Although, the borders of a rectangle are similar to lines, they are bounded by two specified points P and Q.

Without loss of generality, for such a bounded line the direction vector $\vec{u}$ is assumed to be given by these two points according to

$$\vec{u} = \vec{Q} - \vec{P} \quad . \tag{4.14}$$

Additionally, if λ is restricted to the interval

$$0 < \lambda < 1 \quad , \tag{4.15}$$

only the section of the line h between the two points P and Q is referred by the respective vector equation.

An intersection point S of two such bounded lines, e. g., two borders of different rectangles, is calculated, based on the described determining of intersection points of lines in the $\mathbb{R}^2$ vector space. Thereby, for λ_S and λ_S' the restriction

$$0 < \lambda_S < 1 \text{ and } 0 < \lambda_S' < 1 \tag{4.16}$$

is applied, accordingly. Hence, the intersection point S is located on the bounded sector of the lines, respectively.

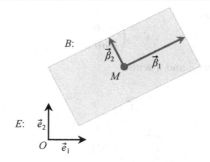

cf. [JH13]

FIGURE 4.6.: Comparison of a basis B defined by a given rectangle in contrast to the common standard orthonormal basis E.

4.3.2. Verifying a Point's Location in Relation to an Area

C2X applications, like the *Weather Hazard Warning*, use arbitrarily oriented rectangles to define areas. Accordingly, in [JH13] the author of this thesis outlines an algorithm to determine, if an arbitrary point, e. g., the vehicles position, is located inside or outside a given geographical area. Moreover, the algorithm is implemented, tested, and verified within the weather related applications in sim[TD] [SIMTD] and DRIVE C2X [DRIVE]. Thus, in this section the respective approach is detailed.

Regarding specifications from Section 2.5.2, each position P is an element of the two dimensional vector space $\mathbb{R}^2$. Hence, the components of the position vector $\vec{P}$ are given according to the standard orthonormal basis E with origin O and the two basis vectors $\vec{e}_1$ and $\vec{e}_2$.

$$E: \quad \vec{O} = \begin{pmatrix} 0 \\ 0 \end{pmatrix} \,, \; \vec{e}_1 = \begin{pmatrix} 1 \\ 0 \end{pmatrix} \,, \; \vec{e}_2 = \begin{pmatrix} 0 \\ 1 \end{pmatrix} \tag{4.17}$$

The basic idea behind the presented method is to transform the components of P into another basis B, whereas the verification is a simple comparison.

An appropriate basis B is derived by the investigated rectangle and its parameters, i. e., M, a, b, and θ, itself. Accordingly, the origin of B is set to the center M of the rectangle, whereas the two basis vectors $\vec{\beta}_1$ and $\vec{\beta}_2$ are oriented in parallel to the borders of the rectangle, each with a length of a and b, respectively. In Figure 4.6 this basis B is depicted in comparison to the standard orthonormal basis E.

While the origin of basis B is given directly by the center point M, the basis vectors $\vec{\beta}_1$ and $\vec{\beta}_2$ have to be calculated with respect to a, b, and θ. Accordingly, based on

θ the vector $\vec{v}$ is given by

$$\vec{v} = \begin{pmatrix} \cos\theta \\ \sin\theta \end{pmatrix} \quad . \tag{4.18}$$

Subsequently, the vectors $\vec{n}$ and $\vec{u}$ with

$$\vec{n} = \frac{\vec{v}}{|\vec{v}|} \tag{4.19}$$

$$\vec{u} = \frac{-2 \cdot a \cdot \begin{pmatrix} -v_2 \\ v_1 \end{pmatrix}}{\left| \begin{pmatrix} -v_2 \\ v_1 \end{pmatrix} \right|} \tag{4.20}$$

are calculated based on $\vec{v}$, respectively.

However, if the rectangle originally is computed as detailed in Section 3.2.3, both vectors $\vec{n}$ and $\vec{u}$ are given and do not have to be determined again.

The two basis vectors of B are determined by just applying

$$\vec{\beta}_1 = \tfrac{1}{2} \cdot \vec{u} \tag{4.21}$$

$$\vec{\beta}_2 = b \cdot \vec{n} \quad , \tag{4.22}$$

leading finally to the basis B as

$$B: \quad \vec{M} = \begin{pmatrix} m_1 \\ m_2 \end{pmatrix} \quad , \quad \vec{\beta}_1 = \begin{pmatrix} \frac{u_1}{2} \\ \frac{u_2}{2} \end{pmatrix} \quad , \quad \vec{\beta}_2 = \begin{pmatrix} b \cdot n_1 \\ b \cdot n_2 \end{pmatrix} \quad . \tag{4.23}$$

However, in general, for an arbitrary point P, the position vector $\vec{P}$ according to basis E is transformed into its equivalent according to basis B by applying the transformation matrix T_B^E. Since this transformation matrix from basis E to basis B is exactly the inverse of the concatenation of the two basis vectors of B, the equation

$$T_B^E = \begin{pmatrix} t_{11} & t_{12} \\ t_{21} & t_{22} \end{pmatrix} = (\beta_1 \beta_2)^{-1} = \begin{pmatrix} b_{11} & b_{12} \\ b_{21} & b_{22} \end{pmatrix}^{-1} \tag{4.24}$$

holds. Consequently, the four components of the 2×2 transformation matrix T_B^E are achieved by solving the linear system of equations

$$\vec{e}_1 = t_{11} \cdot \vec{\beta}_1 + t_{12} \cdot \vec{\beta}_2 \tag{4.25}$$

$$\vec{e}_2 = t_{21} \cdot \vec{\beta}_1 + t_{22} \cdot \vec{\beta}_2 \quad . \tag{4.26}$$

However, in the case of the two dimensional vector space $\mathbb{R}^2$, the solution for this transformation matrix is simply given by

$$T_B^E = \frac{1}{b_{11} \cdot b_{22} - b_{12} \cdot b_{21}} \cdot \begin{pmatrix} b_{22} & -b_{21} \\ -b_{12} & b_{11} \end{pmatrix} \quad . \tag{4.27}$$

Accordingly, each point P, given according to the standard basis E, is transformed into B by multiplying its position vector $\vec{P}$ with T_B^E.

Due to the way basis B is defined, a point P is located inside the rectangle, if the absolute value of each of its components is lower or equal to 1, respectively. Hence, according to basis B

$$|p_1| \leq 1 \quad \text{and} \quad |p_2| \leq 1 \tag{4.28}$$

have to hold. Additionally, P is located exactly on the border of the rectangle if at least one of the absolute values of its components is equal to 1

$$|p_1| = 1 \quad \text{or} \quad |p_2| = 1 \quad . \tag{4.29}$$

Otherwise, the point P is located outside of the rectangle.

4.3.3. Determining the Size of an Overlapping Field

The propagation of two individual situations may overlap only slightly, if they belong to a successive reporting from one vehicle or may overlap heavily, if multiple ITS stations report the same event. Within the Weather Hazard Warning application, this distinction is considered in order to decide whether either the area of an event has to be expanded or if a situation is confirmed by another ITS station. However, the algorithm to define the overlapped field of two rectangles is identical in both mentioned cases.

In [JH13] the author of this thesis proposed an algorithm to determine the size of such an overlapped field, which is detailed in the following. Accordingly, the following steps are performed.

1. All intersection points S_j of each of the four borders of one rectangle with each of the four borders of the other rectangle are calculated.

2. Additionally, all corners E_j of each rectangle, which are located inside the other rectangle, respectively, are determined.

3. Sequentially, the overlapped field is defined by a non-self-intersecting polygon, which vertices are given by the set of all these intersection points and corners if they are ordered the right way.

4. Finally, the area of this polygon can be calculated using the *Gaussian trapezoid formula*.

Such an overlapping field, i. e., a polygon and its vertices, is illustrated in Figure 4.7.

In the first step, the intersection points S_j between the borders of two different rectangles are calculated based on determining intersection points of lines in the $\mathbb{R}^2$ vector space, as detailed in Section 4.3.1.

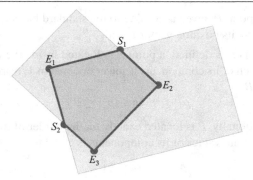

cf. [JH13]

FIGURE 4.7.: The overlapping field, i. e., the area covered by both rectangles.

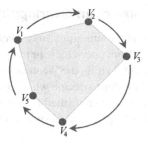

cf. [JH13]

FIGURE 4.8.: A polygon with vertices ordered in turn.

Thus, all intersection points are calculated by cross-checking every border of one rectangle for intersection points with every border of the second rectangle. These checks may either result into

1. an empty set, if the rectangles do not overlap or

2. at least two and up to eight intersection points, if they do overlap.

However, in the first case with non-overlapping rectangles, the overlapping field has no size. Hence, this case is not further regarded.

In Section 4.3.2 a method to determine whether a point is located inside of a rectangle is presented. In the second step, this method is applied to all corners of both rectangles to achieve the corners E_j that are located inside the respective other rectangle. This procedure may result in none up to four points.

Thus, a total number of three up to eight points can be found. These k points are the vertices V_1 to V_k of the overlapping field, if they are ordered accordingly as depicted in Figure 4.8.

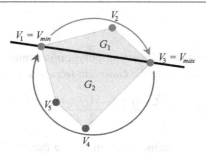

cf. [JH13]

FIGURE 4.9.: Finally ordered vertices.

However, if the points S_j and E_j are collected as described, they are ordered arbitrarily.

The right ordering is achieved by sorting all points according to their respective x-coordinate in the first place. Subsequently, the point V_{min} with the smallest and the point V_{max} with the largest x-coordinate, respectively, define a line. This strait line separates the set of remaining vertices into two groups, G_1 and G_2. Hence, with

$$\vec{n} = \frac{\binom{v_{min,2} - v_{max,2}}{v_{max,1} - v_{min,1}}}{\left| \binom{v_{min,2} - v_{max,2}}{v_{max,1} - v_{min,1}} \right|} \tag{4.30}$$

$$c = v_{max,1} \cdot n_1 + v_{max,2} \cdot n_2 \tag{4.31}$$

for every point V_i a signed distance δ_i is calculated by applying

$$\delta_i = \vec{V}_i \cdot \vec{n} - c \quad . \tag{4.32}$$

Accordingly, each vertex featuring $\delta_i \geq 0$ is located on the one side of the line, so that it belongs to G_1. In contrast, each vertex with $\delta_i < 0$ is located on the other side of the line, therefore it belongs to G_2.

$$G_1 = \{V_i \mid \delta_i \geq 0\} \tag{4.33}$$

$$G_2 = \{V_i \mid \delta_i < 0\} \tag{4.34}$$

Finally, by sorting the elements of both groups according to their x-coordinate, but reversing the sequence of the elements in group G_2, the final order of the vertices V_1 to V_k of the overlapping polygon is given by appending the reversed list G_2 to the end of the list G_1. Such an ordering is depicted in Figure 4.9.

These sorted vertices are surrounding the overlapping field of the two rectangles.

By setting

$$V_{k+1} = V_1 \tag{4.35}$$

the size A_{pol} of this non-self-intersecting polygon is achieved by applying the k calculated vertices $V_j = (x_j, y_j)$ to the *Gaussian trapezoid formula*

$$A_{pol} = \left| \frac{\sum_j^k (x_j + x_{j+1}) \cdot (y_{j+1} - y_j)}{2} \right| . \tag{4.36}$$

This size A_{pol} is an absolute value given in square meters. However, to achieve a percentage of relative overlapping the size A_{pol} has to be set in relation to the size of rectangles. With A_{rect} as the size of, e. g., the larger rectangle, the relative overlapping A_{rel} is given by applying

$$A_{rel} = \frac{A_{pol}}{A_{rect}} \tag{4.37}$$

and, consequently, the relative overlapping in percent by multiplying A_{rel} with 100.

To verify the detailed algorithm and to prove its suitability, it is implemented and tested by the author of this thesis within the *Road Weather Warning* application in sim^{TD} [SIMTD] and, additionally, within the *Weather Warning* application in DRIVE C2X [DRIVE].

4.3.4. Overall Event Area

To determine the overall event area, the included situations, i. e., those who do overlap, have to be regarded. As detailed in Section 3.2.2, situations may include the trace, the respective detecting vehicle has driven. Accordingly, in order to determine the event's area, these traces may be regarded. Hence, the algorithm to determine a rectangle surrounding an arbitrary set of points as detailed in Section 3.2.3 may be applied to the set of all situation's traces.

However, if no trace is available, the situation's edges may be provided to the algorithm, instead. Thus, it is ensured that the event's expanse includes all regarded situation's areas.

Moreover, even if, for what reason ever, neither a trace nor a rectangle is provided, comparable appropriate positions may be passed to the algorithm, instead. Hence, if the situations expanse is represented by a circle or an ellipse, e. g., four to eight equidistantly distributed points located on the border of the area, may be exploited within the algorithm, in exchange, as depicted in Figure 4.10. Hence, with a certain loss of accuracy, an aggregated area might be estimated by applying the presented approach, anyway.

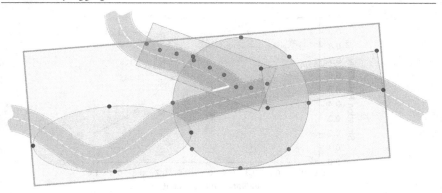

FIGURE 4.10.: Exemplary set of points selected from diverse situation's area representations in order to determine the overall event expanse.

4.4. Reliability Aggregation

In order to aggregate individual situations to an overall weather event, information provided by considered situations have to be merged. This obviously includes the situation's detection reliability assigned by the detecting ITS station. However, the actual reliability of a situation to be present may decrease over time.

Accordingly, within this section an approach to regard detection reliability over time and at situation aggregation is detailed.

4.4.1. Time-dependent Situation Reliability

As detailed in Chapter 3, the *Weather Hazard Warning* application deployed on *Vehicle ITS Stations* or *Central ITS Stations* assigns an initial situation reliability to each detected weather situation. This value is calculated based on the available sensors and their quality. Thus, this detection reliability r_{gen} indicates the probability of the situation to be present at generation time t_{gen} and is included in sent messages as detailed in Section 2.3[45].

However, even if situations expressed by DENMs consists of a maximum validity duration, the probability for the situation to be still present decreases monotonic

---

[45] In Section 2.3 the reliability of a DENM is defined as an *information Quality* level within the range from 0 to a use-case specific maximum value. However, for reasons of simplification and, moreover, unification in context of this section the reliability is regarded as a probability in the range of 0.0 to 1.0. This can easily be archived by dividing the given level by the use-case specific maximum reliability level.

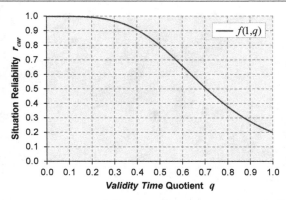

cf. [JH11]

FIGURE 4.11.: Exemplary curve of decreasing situation reliability r_{cur} over past *Validity Time* quotient q with an initial reliability $r_{gen} = 1$ at situation generation time.

over time and is lowest at expire time t_{exp} at the end of the validity duration. Thus, the probability of a situation to be present at current time t_{cur} depends, firstly, on the initial detection reliability r_{gen} assigned on generation time. Additionally, the time quotient q of past situation validity time, given by

$$q = \frac{t_{cur} - t_{gen}}{t_{exp} - t_{gen}} \tag{4.38}$$

has to be regarded.

This decreasing of reliability does not necessarily have to be linear. In fact, if a non-linear function, e. g., as proposed in [JH11], is advocated to calculate the time-dependent reliability, the real decreasing probability is adapted more precisely. Such a non-linear function f is given by

$$r_{cur} = \frac{r_{gen}}{1 + 4 \cdot q^4} \tag{4.39}$$

and its curve is exemplary depicted in Figure 4.11 with respect to an initial reliability of $r_{gen} = 1.0$.

4.4.2. Overall Event Reliability

Since vehicles detect and report weather situations independently from each other, each weather event is announced by multiple origins. On receiver side, this multitude of individual situation reports is aggregated together to an overall event. Consequently, the reliability of the overall event is based on the combination of the individual time-dependent situation reliabilities r_i, which are achieved as detailed in Section 4.4.1.

Within the Weather Hazard Warning application, vehicles work together cooperatively just this way. Hence, a vehicle's detection is confirmed by other vehicles detecting comparable weather situations. Consequently, the event reliability shall be high if the reliability of the single situation is high and, moreover, an event is reported by a high number of vehicles, additionally.

In [HHSV07] an approach to combine multiple reliabilities is stated. Thereby, all n situation's reliability values r_i mapped to one single event are added together.

$$r_{add} = \sum_i^n r_i \qquad (4.40)$$

Obviously, this approach leads to high values, if the single situations have high reliability. Additionally, the reliability increases with the number of situations, which is related to the number of vehicles that report the situation.

Unfortunately, by applying this method an ambiguous situation also leads to high event reliability if the traffic density is high. Thereby, the high amount of vehicles leads to a high amount of detections. Therefore, the individual reliabilities are accumulated to a high overall event reliability even if each situation's reliability is low due to the ambiguous conditions. Thus, by applying this method, even unclear situations achieve a high reliability if there are enough vehicles.

Another way to combine the time-dependent event reliability, is to calculate the mean value of the n individual situation reliabilities r_i.

$$r_{mean} = \frac{\sum_i^n r_i}{n} \qquad (4.41)$$

Obviously, this approach does avoid the problem of accumulating low reliabilities. But, unfortunately, it does not regard increasing event reliability if several vehicles report situations with high reliability, e. g., due to availability of advanced and high quality sensors. Hence, situations with low reliability reported by vehicles which are unable to decide ambiguous conditions are weighted the same as high reliable detections.

Accordingly, while calculating an overall event reliability r_{evt} based on multiple situations from different origins with diverse time-dependent situation reliabilities, individual weights ω_i for each of the n reliabilities r_i have to be advocated.

$$r_{evt} = \frac{\sum_i^n \omega_i \cdot r_i}{\sum_i^n \omega_i} \qquad (4.42)$$

Such a weighting function g providing a weight ω_{cur} to a time-dependent reliability

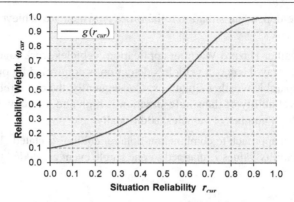

FIGURE 4.12.: Curve of the situation reliability weighting function g, providing a weight ω_{cur} to each time-dependent situation reliability r_{cur}.

r_{cur} with respect to the time-depended reliability itself, is given by

$$\omega_{cur} = \frac{1}{1 + 9 \cdot (1 - r_{cur})^3} \tag{4.43}$$

and the respective curve is depicted in Figure 4.12. This function has similar characteristics as f as introduced in Equation 4.39, but with a fixed first argument, inverted behavior according to the second argument, and a steeper curve.

Finally, a driver notification is triggered only, if the event reliability r_{evt} reaches at least a minimum reliability r_{thr} threshold, as further detailed in Section 5.2.

In contrast to the aggregated overall event area as calculated in Section 4.3, the time-dependent event reliability r_{evt} as introduced in the previous section could not be stored and reused each time it is needed. Instead, r_{evt}, including all underlying situation reliabilities, have to be recalculated each time r_{evt} is requested.

4.5. Situation Management

Within the Weather Hazard Warning application, information about local weather events is maintained by an *Event Manager*. Accordingly, this application's sub-component holds a list of all known and valid events, and their included situations, in the near environment.

Once a weather situation is detected or received within the host vehicle, it is provided to the Event Manager. Within the Event Manager, the new situation is compared against each known event. This comparison leads to three possible results.

Firstly, if no matching event is found, a new event is generated based on the information, e. g., validity duration, reliability, area, and weather type, provided by the new situation. Accordingly, this event contains only the new situation.

Secondly, if there is exactly one matching event, the new situation is integrated into this event. Thereby, the event's information is updated including previous stored data and the new data. Hence, the event's area is updated as detailed in Section 4.3.4 by regarding all included situation's areas. Furthermore, the event's reliability is recalculated as detailed in Section 4.4.2. Additionally, the event's weather type may not be the same as the situation's. Hence, it is updated according to a configurable hierarchy. A possible hierarchy regarding the four weather types *ice on the road*, *heavy rain*, *dense fog*, and *strong winds* is provided in Figure 6.5. Additional weather types may be insert appropriate into the ordering, regarding events severity and probability of one event implying another, such as, e. g., heavy rain most probable implies strong winds.

Finally, if the situation belongs to multiple events, all matching events and the new situation have to be merged, accordingly.

In addition to inserting new situations, known events are maintained by means of periodic *update cycles*. Within an update cycle, each situation within an event is verified to be still valid and removed otherwise. However, if a situation is removed, the event's information has to be recalculated. Thereby, since a situation is removed, it has to be considered that an event may be splited into at least two separated events.

Additionally, during the periodic update cycle within the event manager, it may be verified if the vehicle is approaching the current event. Thus, a respective driver notification can be triggered as further detailed in Chapter 5.

Hence, the Event Manager is a major component of the Weather Hazard Warning application, maintaining almost all gathered information.

5. Driver Notification

The final objective of a road safety related Car-to-X application is to present a notification to the driver of a vehicle, if an event may present a threat to driving safety. This way, the driver is aware of the potential dangerous situation and may adapt the driving behavior, such that possible harm can be avoided. The use-case of the weather Hazard Warning is intended just this way.

As detailed in Chapter 3, each weather event is most likely detected and reported by multiple independent situations, jointly building up the entire propagation of the event. Hence, these individual situations, received as various DENMs are combined together as detailed in Section 4.3. Accordingly, a notification to the driver is presented not for every single situation again and again, but once for an entire event.

In this chapter a method to determine events that are relevant for a presentation, is detailed and discussed in the first place. Thereby, a solution which does not rely on map data is presented. The availability of such data cannot be assumed, at least, within low cost products and during the introduction phase of C2X systems. Subsequently, presentation interfaces, strategies, duration, and intensity are outlined.

5.1. Relevant Events

Car-to-X communication facilitates driver notifications regarding potential dangerous events. Obviously, only events in a use-case specific *"near"* environment have to be regarded.

Thereby, just a few C2X applications, like the *Emergency Vehicle Warning*, consider every event in the environment. For the majority of C2X use-cases, like the *Weather Hazard Warning*, only events located ahead in direction of driving have to be regarded. Hence, in this section, a method to determine such relevant events, i. e., events located ahead and within a maximum distance, is presented.

This approach is based on the current vehicle's distance d_{evt} to the event. Other approaches may regard a *time-to-event* t_{evt}, instead. Since the time-to-event is just calculated based on the distance d_{evt} by regarding, e. g., the vehicles speed v_{ego}

$$t_{evt} = \frac{d_{evt}}{v_{ego}} \quad , \tag{5.1}$$

these approaches may be regarded as equivalent. However, the distance d_{evt} is required, anyway.[46]

5.1.1. Determining the Distance to a Rectangle

As described before, a C2X application may display a notification to the driver while approaching a hazardous event. Thereby, the distance to that event has to be determined to either evaluate whenever the event is relevant for triggering a presentation or, moreover, to display the distance or estimated arrival time. Consequently, in both cases, the distance from the vehicle's current position to the nearest border is calculated.

Thereby, the distance is 0, if the vehicle has already reached the event and is located inside its area. This case may be verified as detailed in Section 4.3.2.

Additionally, only events located in the near environment of the vehicle are of interest for a presentation. Hence, a maximum warning distance D_{max} is assumed. This distance may vary by the respective use-case and, moreover, may be adapted according to further environmental conditions, e. g., the current vehicle speed, as also discussed in Section 5.2.

Finally, since notifications are presented for upcoming events only, the direction of driving ω, has to be considered as well. Thereby, this direction is defined the same way as the rectangle parameter θ. Hence, it is represented as clockwise angle against north.

Determining the distance from an arbitrary point P to the nearest border of a given rectangle, is solved in the vector space $\mathbb{R}^2$. By applying the maximum distance D_{max} and the direction of driving ω, the vector $\vec{w}$ with

$$\vec{w} = D_{max} \cdot \begin{pmatrix} sin\omega \\ cos\omega \end{pmatrix} \tag{5.2}$$

is achieved. Due to its definition, $\vec{w}$ is pointing according into the direction of driving and its length is exactly D_{max}.

Based on the current vehicle's position P and the vector $\vec{w}$ a straight line g is given

---

[46] In complex scenarios, vehicle's and event's speed and acceleration may be regarded to calculate t_{evt}, instead. However, in such cases, same principles by regarding more complex formulas are applied.

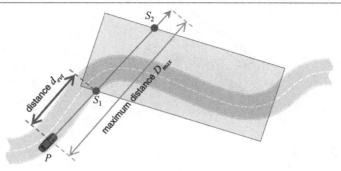

FIGURE 5.1.: Calculation of the distance d_{evt} from a vehicle's current position
to the nearest border of a weather event, regarding the direction of
driving.

by

$$g : x(\lambda) = \vec{P} + \lambda \cdot \vec{w} \quad . \tag{5.3}$$

If λ is bounded to the interval

$$0 < \lambda < 1 \quad , \tag{5.4}$$

the line only refers to the sector starting at P with a length of D_{max}.

According to Section 4.3, each border of a rectangle is given by an according
bounded line h, respectively. Intersection points S_j of all four borders of the
rectangle with the bounded line g are determined, as detailed in Section 4.3.

However, the distance to each border sharing no intersection point with g under the
given constrains can be assumed as infinite.

In contrary, for all determined intersection points S_j, the distance d_j to the position
P is given by

$$d_j = \left| \vec{P} - \vec{S}_j \right| \quad . \tag{5.5}$$

Consequently, from all achieved distances d_j, the minimum is the desired distance
d_{evt} from P to the nearest border of the rectangle, as depicted in Figure 5.1.

5.1.2. The Recognition Field

However, there are cases where this distance calculation will not provide appropriate
results. Assuming the event is located, e. g., right behind a curve. Hence, an
intersection point S_j is not achieved at the desired position P, but not before the
vehicle enters the curve at position P', as depicted in Figure 5.2. For road safety
related C2X use-cases, this might be too late to provide a timely driver notification.

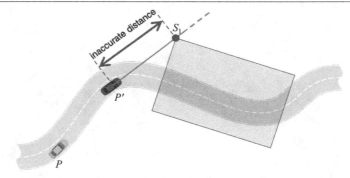

FIGURE 5.2.: The distance from a vehicle's current position to the nearest border of an event is not calculated correctly, if the event is located behind a curve.

Moreover, especially these situations are affected, where C2X communication shall provide the highest advantages compared to local sensors, i. e., if the dangerous event is located behind a curve and out of the line-of-sight.

Accordingly, the distance has to be calculated not only in direct direction of driving, but with an appropriate *opening angle* α. Thereby, the detailed algorithm to achieve intersection points S_j is applied not only the direction of driving ω, but also to selected directions ω' in the range of

$$\omega - \frac{\alpha}{2} \ \leq \ \omega' \ \leq \ \omega + \frac{\alpha}{2} \ . \tag{5.6}$$

However, since the probability of driving into the direction ω' decreases with increasing offset to the current driving direction, the maximum recognition distance D_{max} should by decrease accordingly. This way, a *recognition field* in front of the vehicle is created, as depicted in Figure 5.3. Thereby, not only events located directly ahead are regarded, but also those that are inside the specified opening angle.

The actual values for D_{max} and α may vary not only according to the C2X use-case, but also according to the current driving speed, current road class, or the driver attentiveness, as further detailed in Section 5.2.

5.2. Presenting Driver Notifications

How and when driver notifications are presented has a huge influence on the effect these notifications have. Thereby, the appropriate notification design, interfaces, and

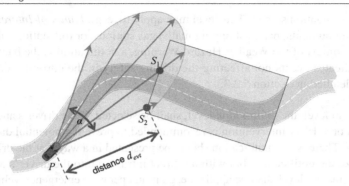

FIGURE 5.3.: Calculation of the distance d_{evt} from a vehicle's current position to the nearest border of a weather event, with regarding an opening angle α and decreasing maximum recognition distance D_{max}.

strategies affect the driver's behavior as well as the information itself. Accordingly, various psychological aspects according to human perception and behavior have to be considered [POR11]. Consequently, within this thesis a short introduction and overview is stated.

Additionally, for C2X communication, depending on the criticality of the current scenario, a suitable driver notification with adequate intensity shall be selected. Hence, within this section appropriate notification levels for C2X use-cases are defined and according interfaces for presenting a notification and catching the driver's attention are discussed. Finally, exemplary strategies and instructions for determining and selecting suitable notifications within Weather Hazard Warning scenarios are detailed.

5.2.1. Notification Levels

For driver notifications in C2X communication scenarios within the context of this thesis the author defines four *notification levels* with increasing priority:

1. *background*,
2. *information*,
3. *warning*, and
4. *severe*.

The lowest notification level, i. e., the *background* level, may be applied for C2X events with informational purpose only, such that no driver intervention is required. A notification thereby may be presented in a very decent way, e. g., as an icon within a navigation screen. This level may apply to, e. g., *Points-of-Interests* (PoI) such as monuments, parks, shopping malls, gas stations, or toll stations along the road. In context of the Weather Hazard Warning, a notification at the background level indicates events not affecting the driving safety in the current scenario, as further detailed in Section 5.2.3.

The second level, the *information* level, shall be selected, if the driver's attention is required or a driver intervention is recommended to prevent a potential dangerous situation. Thereby, the notification should be presented in a way that the driver may recognize the notification, but without direct intrusion into the driver's attention. The information level shall be applied, e. g., if an operating emergency vehicle is in the near environment such that it has to be considered or if a wrecked vehicle is located at the road side such that special attention is required. The Weather Hazard Warning exploits the information level if an event may affect the driving safety in the current scenario, as further detailed in Section 5.2.3.

The *warning* level is the first intensive notification level. It shall be exploited if a driver intervention is necessarily required to avoid damage or a highly dangerous situation. Consequently, the notification shall be presented in a way that the driver may directly recognize the notification such that adequate actions may be timely applied. Such a warning level shall be used, e. g., if an operating emergency vehicle is blocked, if traffic regulations are violated, or if uncomfortable hard braking is required to avoid a collision. Within the Weather Hazard Warning application, a driver notification at the warning level is selected if the current driving dynamics are not adapted to the current or upcoming weather situation, such that the road safety will not be ensured anymore, as further detailed in Section 5.2.3.

The highest notification level, the *severe* level, is reserved for scenarios, where immediate and intense driver intervention is required to avoid damage or, at least, to mitigate the resulting damage if it is unavoidable at all. Accordingly, the notification has to be intense and implying immediate driver reaction. Moreover, the notification itself shall preferably affect and initiate the reaction, without causing the driver to decide and select an action by its own, as further outlined in Section 5.2.2. Due to its intensity, this level applies only to highly dangerous situations, e. g., to hard braking or intersection situation, where a rear- or side-impact crash may be prevented only with a full brake or where even a full brake may only mitigate the impact. According to the opinion of the author of this thesis, such an intensive notification is not applicable for weather caused situations and, hence, not applied by the Weather Hazard Warning.

5.2.2. Driver Notification Interfaces

Within recent C2X communication research projects and standardization, driver notification interfaces focus on visual approaches. Thereby, icons and short text messages are displayed on the navigation or a comparable screen or mobile device, as also presented in Section 6.3 and Section 6.4. This visual indication is supported by acoustic signals at higher notification levels [SIMTD09d] [DRIVE11c].

Due to their size and resolution, such navigation or comparable screens are highly appropriate for displaying icons indicating, e. g., *Points-of-Interests* (PoI) along the road, even for a larger scale above multiple kilometers. Moreover, they are capable to display additional textual information regarding upcoming C2X related events or PoIs. Hence, driver notifications at background level and additional information regarding higher prioritized notification, e. g., the distance to the event, may be display on such a screen.

However, since such navigation or comparable screens are not in the driver's direct field of view, they are not applicable for notifications at higher levels. Consequently, a notification displayed in the instrument panel is more appropriate for notifications at the information, warning, and severe level.

Due to the limited space available within most instrument panels available in today's vehicles, only a significant icon indicating the upcoming event may be displayed. However, driver's reaction on *icon only* approaches is significant faster than on *text, icon with text*, or *speech*-based notifications [CCMM09]. Nevertheless, additional information may be displayed on a supplementary screen, e. g., the navigation screen.

The icon placed in the instrument panel may be complemented only by the distance to the event for notifications at the information and warning level. Such additional information provided to the driver is mandatory for timely and efficient reactions. Since notifications at the severe level imply immediate driver reaction, no additional information shall be provided.

However, such icons may be noticed too late for notifications at warning and severe level. Hence, supplementing techniques to focus the driver's attention are required. Consequently, notifications at higher levels may be accompanied by acoustic signals or vibrations within the steering wheel or seats.

Additionally, an LED stripe placed around the front shield may visually indicate the danger and, hence, focus the driver's attention. Moreover, this approach facilitates indication of the direction, where the event is located.

An approach based on a *Head-up Display* (HUD) projected onto the frond shield would combine all these advantages. It is projected onto the frond shield, such that it is located perfectly in the driver's direct field of view. Additionally, indication of directions or even highlighting individual objects is facilitated. Such advanced displaying capabilities catching up *Augmented Reality* (AR) may be applied for notifications at background, information, warning, and severe level. Thereby, icon sizes, color and intensity shall be adapted to the notifications priority.

Additionally, providing *action suggestions* to the driver enhances the results [CMC+10]. Thus, suggestions may be visually included as far as possible into an instrument panel based solution or a HUD. Moreover, a force-feedback indicating and influencing the required action may be exploited for notifications at warning and severe level. Thus, e. g., an contra force on the throttle may be applied, if a deceleration is required or a force on the steering wheel suggests a turning maneuver. However, for severe notifications also smooth driving interventions such as, e. g., pre-braking, may be applied.

5.2.3. Weather Hazard Warning Notifications Strategies

Driver notifications within the Weather Hazard Warning application intend to timely inform the driver about dangerous situations that need special attention. Thus, an appropriate driving behavior regarding environmental conditions can be achieved. Hence, to ensure a timely notification, a presentation is triggered if the vehicle is approaching a weather event and the distance to the event d_{evt} or the time to the event t_{evt} is below respective thresholds D_{max} or T_{max}.

To achieve high-quality assistance, *Advanced Driver Assistant Systems* (ADAS), like the Weather Hazard Warning application as presented in this thesis, shall be adaptive to the driver's current driving behavior, e. g., speed, and state, i. e., workload and attentiveness [CTM10]. Consequently, in order to still ensure a timely notification, actual values for D_{max} or T_{max} are increased with driving speed and decreasing driver's attentiveness. This way, a presentation is guaranteed to appear for an adequate period to adjust driving behavior calmly.

Since driver's capability and necessity to adapt driving behavior depend on the actual road class, i. e., highway, rural roads, or urban roads, it is considered in addition. Since a rural road may contain more and sharper curves as a comparable strait highway, regarding, e. g., slippery roads or reduced visibility due to fog, the possible driving speed on a highway is above the possible driving speed on a rural road, .

Moreover, the fuzziness of the most probable driving path depends on both, the driving speed and the road class. Thus, the parameter α determining the shape of the *recognition field* as presented in Section 5.1.2, is adaptive on speed and road class. Consequently, since the most probable path is more determined on highways than on rural or urban roads, the opening angle α is decreased on highways.

Based on the actual determined thresholds D_{max} and α, the Weather Hazard Warning notification strategy includes a first notification if the event is located far away. Since there is no urgent driver reaction required at this early state, this notification is presented at *background* level.

However, in accordance with D_{max} and α, the selected *notification level* as presented in Section 5.2.1 is adaptive, too. Hence, for subsequent presentations during approaching an event the driver's behavior, attentiveness, and reaction, is considered.

If, e. g., the driving speed, curve speed, and braking intensity, are already below respective thresholds, accurate driving maneuvers regarding the current weather situation, are assumed. Hence, no adaption of the driving behavior is required and, thus, no more conspicuous notification at *information* or even *warning* level is presented.

In contrast, if the event is approaching and driving maneuvers are not suitable regarding the current weather situation, the notification level is increased to *information*.

However, after presenting the notification at information level and reaching a distance below a threshold D_{warn}, insufficient or no adaption at all, may be recognized. In this case, catching driver's attention becomes more essential. Thus, a more conspicuous driver notification at *warning* level is selected. As detailed in Section 5.2.1, this may include not only visual but acoustic and haptic interfaces, too.

Anyway, if the driver adequately adapts its driving behavior the notification level is lowered to information or background level, accordingly.

Finally, since the driver may recognize the event by its own eyes, in most cases the notification is removed if the vehicle reaches the event. However, even if an event is already reached, heavily inappropriate driving maneuvers may trigger a presentation again.

In contrast, slippery road or strong cross winds are hard to be recognized by a driver. Hence, for such events, a notification at background level is presented as long as the vehicle is inside the affected area.

The multiple mentioned points, affecting the notification level are summarized in Figure 5.4.

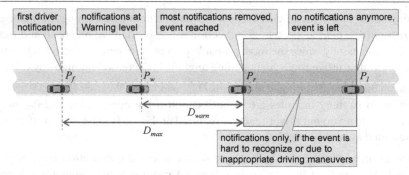

FIGURE 5.4.: At P_f the first notification appears, beyond P_w also notifications at
warning level may be triggered, at P_r most of the notifications may
disappear since the event is reached, and P_l, where the event is left
and no notification appears anymore.

Besides notification distance and level, also the reliability of an event r_{evt} as detailed
in Section 4.4.2 has to be considered when triggering driver notifications. Hence, in
order to avoid annoying false-positive notifications, events with an overall reliability
level below an appropriate threshold r_{info} are displayed at *background* level, only.
Accordingly, events with a reliability below the threshold r_{warn} are not displayed at
warning level. Thereby, for r_{info} and r_{warn}

$$0 \leq r_{info} \leq r_{warn} \leq 1 \qquad (5.7)$$

holds.

However, it still could not be guaranteed to filter false-positive detections originated
from single vehicles by this reliability based approach. Hence, a complementary
procedure has to be applied, additionally, before a driver notification is triggered.

Therefore, the number n of vehicles reporting an event and the number m of vehicles
passing the event's location without reporting is considered. Thus, the ratio n_{ratio}

$$n_{ratio} = \frac{n}{n+m} \quad , \qquad (5.8)$$

is regarded on selecting the level of a notification as follows:

if $0.0 \leq n_{ratio} < 0.5$: no notification is displayed at all,

if $0.5 \leq n_{ratio} < 0.9$: notifications are displayed only at information level, and

if $0.9 \leq n_{ratio} < 1.0$: all notification are displayed according to their priority.

Obviously, this includes only vehicles capable to detect weather situations and is not applicable for slippery road events, since slippery road detection highly depends on the current driving maneuver.

Additionally, the number n of vehicles reporting an event is considered. Hence, if an event is not reported by a least n_{thr} different origins, the overall event reliability r_{evt} is set to -1. Accordingly, the reliability regarded to trigger a driver notification r_{evt} is given by

$$r_{evt} = \begin{cases} r_{evt} & \text{if} \quad n \geq n_{thr} \\ -1 & \text{else} \end{cases} \qquad (5.9)$$

Hence, false-positives from single origins may be of no consequence and, thus, false notifications are avoided.

6. Reference Implementations

Within this chapter realizations of concepts and components presented in previous chapters are detailed and discussed. This includes an overall C2X system architecture and an implementation and integration of the Mobility Data Verification component. Finally, two field operational trial implementations of the Weather Hazard Warning application and their integration into the respective C2X system architectures are detailed.[47]

However, for every implementation slight adaptions of the basic concept have to be applied, in order to concern environmental restrictions. These modifications are detailed in the respective sections. The presented implementations are consulted for an extensive evaluation as presented in Chapter 7.

6.1. The sim^{TD} Field Operational Test System Architecture

To examine the usability and the impact of C2X use-cases to traffic efficiency and road safety, research projects are established. Within such projects, C2X system architectures are developed, use-cases and applications implemented, and real world test performed. An exemplary research project's system architecture is detailed in this section. Thus, an in-depth view on a prototypical near series C2X system is provided.

Within the German research project *Safe and Intelligent Mobility – Test Field Germany* (sim^{TD}) leading car manufactures, automotive suppliers, public authorities, and research institutes join together to build up a large scale field operational trial for Car-to-X communication [SIMTD]. Thereby, almost the entire spectrum of use-cases facilitated by C2X communication is covered. Hence, more than 20

---

[47] In addition, under supervision of the author of this thesis, a reduced third implementation for a *lightweight* C2X system [BER11] is developed in [JÄN11]. However, since this implementation has to be heavily adapted in order to regard system boundaries, this approach is not further detailed in this thesis.

applications, including a Weather Hazard Warning as further detailed in Section 6.3, are implemented and evaluated.[48]

To perform realistic real world Car-to-Car communication scenarios, more than 100 vehicles are equipped with C2X communication systems. Moreover, in order to include Infrastructure-to-Car and Car-to-Infrastructure communication, more than 100 roadside stations are build up along highways, rural roads, and urban roads in and around the city of Frankfurt am Main, Germany. Additionally, scenarios which are hardly realizable within real word trials, e. g., large scale traffic flow scenarios, are covered by traffic simulation. Furthermore, e. g., possible dangerous or driver acceptance related scenarios are examined within driving simulation as further detailed in Section 7.5.

To facilitate the large scale field operational trial located in the Rhein-Main metropolitan area, a novel C2X system architecture is co-developed by the author of this thesis. Thereby, already existing C2X system architectures designed or developed, e. g., during standardization or within previous field operational trials, are regarded [SIMTD09c] within a basic system concept [SIMTD09f]. Additionally, requests demanded by the multitude of C2X applications [SIMTD09a] as well as requirements for measuring [SIMTD12b], evaluating [SIMTD09n], and controlling the field operational tests [SIMTD10c] are considered within the overall sim^{TD} system architecture [SIMTD09g].

This system architecture is designed as reference for future C2X system development. A short overview of the overall sim^{TD} system architecture is given in [SBH$^+$10], whereas a detailed description is available in [SIMTD10b]. Additionally, Section 6.1.1 summarizes an overview of the system architecture and provides a deeper view on the main components and basic interconnect within the C2X system. Accordingly, an overview of the architecture of a near series C2X vehicle station and its main components is provided in Section 6.1.2. Aligned with the context of this thesis, especially parts related to the Weather Hazard Warning application within sim^{TD} are described in these sections.

6.1.1. Overall System Architecture Overview

The overall field operational test system architecture of sim^{TD} includes all three station types, i. e., *Vehicle ITS Stations* (VIS), *Roadside ITS Stations* (RIS), and *Central ITS Stations* (CIS). Thus, a wide spectrum of C2X communication use-cases is covered.

---

[48] A complete list and description of all sim^{TD} applications is provided in [SIMTD09b].

As a Central ITS Station, the *Integrierte Gesamtleit Zentrale* (IGLZ) of the City of Frankfurt am Main is connected to sim^{TD}. The IGLZ, beside others, observes urban traffic, parking garage usage, and controls adaptive traffic lights within the field operational test area. Accordingly, Roadside ITS Stations located in the urban area of Frankfurt am Main, are directly connected to and partially controlled by the IGLZ. These RIS, if located at a traffic light, provide information about the current traffic light phase and its duration. Moreover, these stations are capable to control the traffic light's phases, if needed within a special field operational test scenario and enabled by the IGLZ.

On the other hand, the *Hessian Traffic Center* (HTC) is consulted as a second Central ITS Station. The HTC, beside others, observes highway traffic and controls electronic traffic signs, e. g., dynamic speed limits or traffic information, along the highways within the field operational test area. Accordingly, Roadside ITS Stations located along the highways of the Rhein-Main metropolitan area, are associated to the HTC, but connected to and partially controlled by the *Test Management Center* (TMC), These RIS serve as gateway to distribute information provided by the HTC.

The TMC is located at the HTC and is established to observe and control the field operational trials. It is connected by wire to all CIS and RIS[49], as well as by mobile telephony to all VIS within sim^{TD}. However, this part of the system architecture is field operational test related, only, and, hence, not deeply detailed.

Beside the traffic related information provided by the two CISs, the concept of C2X communication foresees additional infrastructure data. Thus, road weather related data provided by the *Straßenwetter-Informationssystem* (SWIS) of the *Deutscher Wetterdienst* (DWD) are provided to sim^{TD} by the HTC. Hence, within the field operational test area, data from six highly sophisticated and well equipped read weather measurement stations, as listed in Table 6.1, are available to the Weather Hazard Warning application.

Within C2X communication vehicles and roadside stations are interconnected by wireless communication according to IEEE 802.11p [IEEE10]. Accordingly within sim^{TD} all RIS and VIS are equipped with respective communication devices. Additionally, to evaluate other communication technologies, also common wireless communication according to IEEE 802.11b/g is available.

Moreover, a third communication link based on mobile telephony is available on vehicles. On the one hand, this link is established for test management and vehicle coordination. Since within the Rhein-Main metropolitan area mobile telephony

---

[49] In some cases, if no wired connection is available to a dedicated RIS location, the connection is tunneled within a mobile telephony interface as a fall back solution.

Road	Location	Coordinates	
A5	Europabrücke, Frankfurt	50° 5′ 29.16″ N	8° 37′ 1.11″ E
A5	Motorway Station Wetterau	50° 21′ 12.84″ N	8° 41′ 35.63″ E
A648	Opelrondell, Frankfurt	50° 6′ 57.07″ N	8° 37′ 43.01″ E
A661	Frankfurt-Seckbach	50° 8′ 57.51″ N	8° 42′ 15.88″ E
B40	Mainbrücke, Sindlingen	50° 4′ 31.34″ N	8° 31′ 22.94″ E
B456	Saalburg, Bad Homburg	50° 16′ 11.11″ N	8° 34′ 7.89″ E

TABLE 6.1.: Associated road and location of the six weather measurement stations available within the field operational trial of sim^{TD}.

is available at almost every location, a throughout observation of and connection to vehicles is facilitated. However, beside this field operational test related aspect, within sim^{TD} also the suitability of providing selected information via mobile telephony is evaluated.

As proposed in [C2C07], Vehicle ITS Stations as well as Roadside ITS Stations are composed out of two separated units, each. All communication-related tasks from medium access up to network and facility layer, as well as collection of vehicle sensor data, are performed by the *Communication Control Unit* (CCU). Accordingly, this unit is connection to the C2X antenna. In sim^{TD}, the antenna combines all three communication technologies as well as a GPS antenna within one device. In contrast, the *Application Unit* (AU) hosts all road safety and traffic efficiency related C2X applications, the HMI connection, the field operational trial logging, and the navigation.

Additionally, all vehicles are equipped with a *Human Machine Interface* (HMI), an automotive capable TFT touch-screen with speaker, attached to the AU. This interface is selected to present driver information and warning and may collect driver inputs, both requested by C2X applications or the TMC, respectively.

A simplified overview of the sim^{TD} overall system architecture and interconnection of main stations and components is depicted in Figure 6.1, whereas a more detailed overview of the vehicle station is described in Section 6.1.2.

6.1.2. Vehicle ITS Station Architecture

To facilitate C2X communication, Vehicle ITS Stations are deployed within vehicles. This section provides an overview of the *sim^{TD} Vehicle ITS Station* and its main components.[50]

---

[50] The architecture and components of the *sim^{TD} Roadside ITS Station* differs from the vehicle station in a few details, only. Main differences are related, e. g., to positioning or external interfaces. The

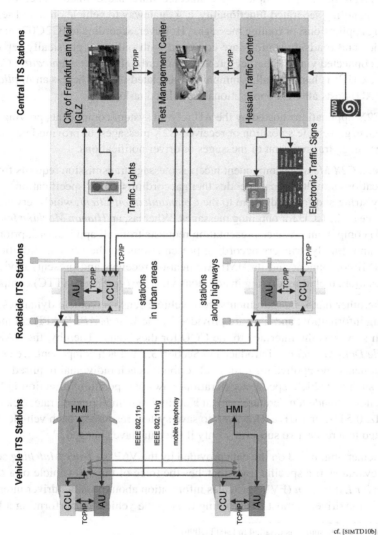

cf. [SIMTD10b]

FIGURE 6.1.: Summary of the overall sim^{TD}architecture with simplified composition and interconnection of main stations and components.

For designing a VIS and RIS multiple approaches have been developed so far. In [ETSI10a], e. g., multiple interconnected units are defined. Thereby, each unit is handling separated functionality, e. g., gateway to vehicle on-board sensors, hosting applications, or routing messages. However, according to [C2C07] in sim^{TD} vehicles and roadside stations are equipped with only two physically separated units, connected via an Ethernet interface. Accordingly, a *Communication Control Unit* (CCU)[51] is handling all communication-related tasks, whereas an *Application Unit* (AU) hosts all C2X applications and related tasks.

The C2X applications hosted by the AU rely on system components proving information, e. g., vehicle's position or received C2X messages, or proving interfaces to request, e. g., transmission of messages or driver notifications.

Hence, a *C2X Message* component receives message transmission requests from all applications simultaneous, encodes them according to their specification[52] into a binary string and forwards them to the *Communication Client*, which serves as the interface to the CCU for outgoing messages. Whereas, an *Human Machine Interface* (HMI) component receives presentation requests from all applications, prioritizes them and, finally, triggers according presentations on the screen. Additionally, the HMI component and its HMI-Application receives and presents navigation information, including driving instructions via a *Text-to-Speech* (TTC) component.

On the other hand, measurements from vehicle sensors, vehicle dynamics, positioning information, and time are provided by the *Vehicle API* (VAPI) component, which serves as the interface to the CCU for these data. Thereby, the VAPI is a *Vehicle Data Provider* as introduced in Section 3.1.1. Each component, i. e., system components and applications, may subscribe for each individual required vehicle data, such as vehicle speed, acceleration, yaw-rate, position, direction indicator activation, outside temperature, etc., at a desired maximum update rate, i. e., one of 0.1 Hz, 0.5 Hz, or 1.0 Hz. However, to save system resources, each vehicle data is updated to a respective subscriber only if the data have changed.[53]

Supplementing, based on the data provided by the VAPI, a *Trace Matching* component evaluates if a specific trace matches the trace of the ego vehicle or a *Driver Behavior Estimation* (FVV) provides information about possible driver intentions, e. g., if the driver is most likely going to stop the vehicle or performing a left or

---

sim^{TD} roadside station is detailed in [SIMTD09h].

[51] Within [C2C07] this unit is referred to as *On-Board Unit* (OBU)

[52] In sim^{TD}, the *Abstract Syntax Notation One* (ASN.1) [ITU09a] is exploited to specify and encode [ITU09b] C2X messages [SIMTD09j].

[53] More comprehensive description of the mechanisms of the VAPI are detailed, e. g., in [EK06], where the VAPI exploited in sim^{TD} is based on.

right turn. Additionally, based on stored map data within a *Navigation SDK*, the *Navigation and Map* component provides information about, e. g., the class of the current road, i. e., highway, rural, or urban.

All incoming messages are provided by the *Communication Client*, which serves as interface to the CCU for incoming messages, to the local message container. This *Environment Table* component stores messages as long as they are valid. Additionally, all messages are evaluated by the *Plausibility Verification* component, i. e., the Mobility Data Verification at as presented in Section 4.1 and further detailed in Section 6.2. However, an individual C2X application usually does not rely on all incoming C2X messages. In contrast, according to the implemented use-case only messages, e. g., originated within a specified distance, direction, or time or with a special cause code are needed. Thus, an application configures the *Relevance Filter* component to provide only these messages from the message container to the individual application.

Since sim^TD is a field operational test, in addition to these C2X communication related system components, test related components are available, too. During develop and test of C2X applications and system components, a *Trace Player* component may provide recorded vehicle data and position to the VAPI, such that the AU may run applications and system components even if it is currently not integrated in a vehicle. During execution of field operational trials, in order to manage the vehicles and tests, the *Test Control* component provides navigation destinations to the Navigation and Map component as well as driver instructions to the HMI component. Additionally, current vehicle position is transmitted to the TMC and driver feedback on use-cases is gathered. Finally, to evaluate the field operational trials, a *Logging* component receives logging data from almost all system components and applications within the AU and stores them on the hard disk.

Within, sim^TD all C2X applications and main system components are implemented as individual *Java-OSGi-bundles*. They are integrated into an OSGi-Framework running within the Java-VM on the AU. However, several components, i. e., the *HMI-Application*, the *Text-to-Speech* (TTC) library, and the *Navigation SDK*, are executed outside the Java-VM and connected via a TCP-Port or the *Java Native Interface* (JNI), respectively.

The sim^TD AU is connected to the second unit, the sim^TD CCU via a wired Ethernet connection. Through this link, all vehicle data are forwarded from the CCU to the AU as well as all C2X messages are exchanged between both units in respective direction.

The major tasks performed by the *Communication Control Unit* are providing vehicle data to the AU as well as receiving and transmitting C2X messages. In contrast to the AU, components hosted by the CCU are executed directly within the operating system.

Within the CCU the *VAPI Server* forwards available vehicle data and position information to the VAPI component hosted by the AU. Moreover, according to the VAPI placed at the AU, the VAPI Server provides vehicle data to other components hosted by the CCU.

Within the vehicle, data are distributed within one or multiple CAN busses. Accordingly, the CCU may be connected up to three busses, which are accessed by the *Low Level CAN Framework* (LLCF) included in the Linux kernel.[54]

Available vehicle data, their resolution, units, update frequency, etc., vary heavily between different vehicles. However, the VAPI Server relies on specified data elements and formats. Hence, a vehicle specific component, the *Vehicle Profile*, converts vehicle specific CAN data provided by the LLCF into the specified elements and formats needed by the VAPI Server. In addition, most vehicle data are available and updated within the vehicles CAN every 10 ms, i. e., at a high rate of 0.01 Hz. Those data are throttle to a frequency of 0.1 Hz, i. e., 100 ms, by the Vehicle Profile.

Positioning information is provided by the *Global Positioning System* (GPS) to the CCU. Those raw positioning information is enhanced with Differential GPS (DGPS) to, e. g., eliminate inaccuracies caused by atmospheric disturbances. However, further smoothing and interpolation of the position is done by the *Improved Positioning* component. Moreover, this component estimates the vehicle's position in cases of short periods of GPS signal loss, e. g., in tunnels, by consulting vehicle data, such as speed, acceleration, steering wheel torque, etc. This position smoothing, interpolation, and *dead reckoning* relies, on one side, on dynamic vehicle data provided by the VAPI Server as well as on static vehicle characteristics, like width, length, height, weight, center of gravity, wheelbase, etc., provided by the Vehicle Profile, on the other side.

In addition, a precise system time within CCU and AU is applied by exploiting the time information included in the GPS signal to apply the *Network Time Protocol* (NTP). Hence, a synchronized time base between vehicles, with derivations in the range from 1 – 4 ms, is achieved. This way, C2X applications may highly trust in timing information provided by C2X messages.

---

[54] Meanwhile, the LLCF project is renamed to *SocketCAN* [SOCAN]. A detailed description of the usage of the framework and its lower layer access is provided by [KB12] and [HAR12].

The gateway to the AU for C2X communication is the *Wireless Manager*, which handles all incoming and outgoing messages. It manages and controls the three communication links, i. e., IEEE 802.11p, IEEE 802.11b/g, and mobile telephony, available within sim^{TD}, in addition.

For all IT-security related tasks the Wireless Manager is directly connected to the *Security Daemon* component. The tasks performed by the daemon include especially message signing and verification as detailed in Section A.3 as well as triggering of pseudonym changes as detailed in Section 2.4.[55]

To facilitate communication within the major communication link, i. e., C2X communication within *ITS-G5A* band at 5.9 GHz based on IEEE 802.11p, the Wireless Manager relies on components implementing the dedicated C2X communication protocol. Thereby, the complete C2X communication stack, from *C2X Facilities*, over *C2X Network*, down to the *C2X Access*, is provided by respective components. Hence, C2X message frequency control and C2X forwarding mechanism as detailed in Section 2.3.4 and, moreover, the complete generation and frequency adaption of CAMs, as detailed in Section 2.3.1, are performed by these components. Thereby, vehicle data required for C2X networking are forwarded via the Wireless Manager. In addition, data such as, e. g., vehicle's position, heading, and speed, are directly included by the lower layers into C2X messages.

In contrast, the two other communication links, i. e., firstly, common wireless communication at 2.4 GHz based on IEEE 802.11b/g, i. e., *WiFi*, and, secondly, communication based on mobile telephony, i. e., *Universal Mobile Telecommunications System* (UMTS), are based on TCP, UDP, and IP. Hence, the Wireless Manager relies on a component implementing the respective stacks. C2X messages and test control related massages are wrapped by these communication protocols.

Finally, to receive and transmit messages within all three communication links, the CCU is connection to the sim^{TD} antenna module providing the physical access. This antenna module includes one WiFi and one UMTS antenna to facilitate common communication technologies. Supplementing, a GPS antenna is included to provide GPS positioning information to the CCU. Completing, to facilitate C2X communication, the antenna module includes even two antennas for transmission and reception at 5.9 GHz. The radiation pattern of one of the antennas is oriented to the front, whereas the radiation pattern of the second antenna is oriented to the back. This way, reliable communication at the ITS-G5A band is facilitated within front and back direction of the vehicle.

---

[55] Further details on the Security Daemon are provided by [GF09], where the sim^{TD} component is based on.

In addition to these C2X communication related components, according to the AU, a *Logging* component is places on the CCU, too. This component collects individual logs created by the multiple CCU components and forwards them to the AU where they are stored to facilitate evaluation of the field operational tests.

In Figure 6.2 a simplified overview of the sim^{TD} Vehicle ITS Station, i. e., Application Unit and Communication and Control Unit with their main components and interconnections, is depicted.

6.2. The Plausibility Verification Component

As further detailed in Appendix A, security based on cryptography is not sufficient to facilitate reliable C2X communication. This is not only due to avoid possible attacks, but also to filter broken sensor values. Hence, the *Mobility Data Verification* is introduced in Section 4.1. Thereby, mobility information, i. e., position, speed, and heading, included in every CAM and DENM, as detailed in Section 2.3, are evaluated to be in accordance with common vehicle dynamics.

In order to secure the sim^{TD} field operational tests and to improve message reliability within the research project, mobility data verification is included into the sim^{TD} field operational trial system architecture. Hence, an adaption of the detailed Mobility Data Verification component is co-implemented by the author of this thesis and placed into the sim^{TD} Application Unit as the system component *Plausibility Verification*, as outlined in Section 6.1.2.[56]

Due to restrictions by the large scale field operational trial, this adapted version does not include the *Local Sensor Check*. However, Section 6.2.2 details a prototypical approach to include such a *local sensor data fusion* into the sim^{TD} architecture, anyhow.

As the majority of the sim^{TD} components located within the Application Unit, the *Plausibility Verification* component is implemented and integrated as individual Java/OSGi-bundle within the OSGi-framework. Hence, it provides services to and registers for services provided by other components.

Thereby, the component subscribes for the ego vehicle position at the VAPI system component, as it is required to do the *Threshold Checks* as well as the *Margin Check*. On the other sides, the Plausibility Verification, of course, provides a service to verify the mobility data included in a C2X message. This service is used for every incoming message by the *Communication Client* component. Within the

---

[56] The integration of the Plausibility Verification is detailed in the respective project documentation at [SIMTD09g], [SIMTD09i], [SIMTD10a], and [SIMTD10b].

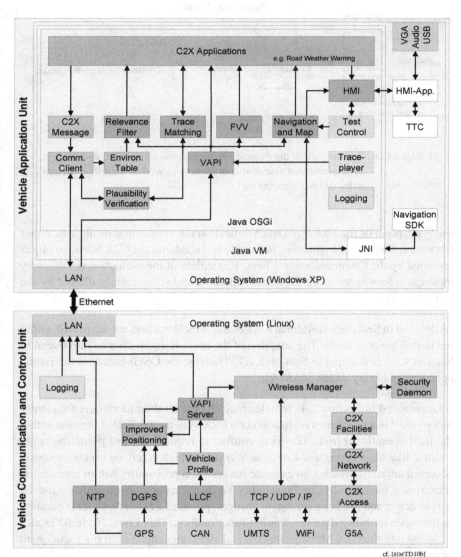

FIGURE 6.2.: The main sim^TD components with simplified interconnections and their distribution onto vehicle *Application Unit* and vehicle *Communication and Control Unit*.

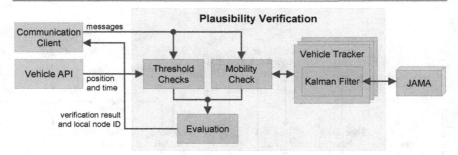

FIGURE 6.3.: Architecture of the *Plausibility Verification* component, with respect
to implemented essential components as well as their internal and
external interconnections.

sim^{TD} adaption of the Mobility Data Verification the verification result, i. e., either
Approved, Neutral, or Erroneous, is directly included into the C2X Message object
provided by the Communication Client. Regardless of the evaluation result, every
message is forwarded to the Environment Table and subsequently filtered by the
Relevance Filter.

As detailed in Section 4.1, Kalman filter-based vehicle trackers are advocated within
the mobility verification. The adaption of the basic Kalman filter as introduced in
Section 4.1.1 is detailed in Section 6.2.1. Thereby, the OSGi-bundle relies on the
open source library *JAMA* [JAMA].

As mentioned in Section 2.4, vehicles may change all their identifiers frequently.
Since sim^{TD} is a near series realization of a C2X system, this is also the case within
the field operational trial. This is in conflict to requirements of particular applications, like the *Intersection Collision Warning*, which is relying on continuously
observed adjacent vehicles, to estimate the collision probability within intersection
scenarios. Consequently, the Plausibility Verification component, which is indeed
able to detect and resolve pseudonym changes, provides a *Local Node ID* in addition
to the basic mobility data verification functionality. This Local Node ID is also
included into the C2X Message objects and stays unchanged even if a pseudonym
change occurs.

The architecture and integration of the *Plausibility Verification* component into
the sim^{TD} AU architecture by means of major components in interconnections is
depicted in Figure 6.3.

6.2.1. Adapting the Kalman Filter as Vehicle Tracker

To evaluate the trustworthiness of received vehicle mobility information, the data are verified to be in consistency to previously received information. Hence, it is necessary to follow the movement of each vehicle. A Kalman filter is appropriate for this purpose [BP99] and, therefore, exploited as vehicle tracker in the presented Mobility Data Verification component.

As already mentioned, C2X messages such as CAMs and DENMs contain mobility information, i. e., position, speed, and heading. According to [ETSI11b] and [ETSI13c] every position is given as WGS-84 coordinate, speed is given in meters per second, and heading is given in degrees clockwise from north. However, exploiting these mobility data directly as system state and measurement vectors for a Kalman filter will result in rather complex matrix calculations.

Hence, for reasons of efficiency, positions are converted into a Cartesian coordinate system, as detailed in Section 2.5.2. Thus, a two dimensional plane with an orthonormal basis is provided, wherein coordinates are given according to an x- and a y-axis, and distances are calculated in meters. Accordingly, speed and heading are combined together and transformed into speed related to each of these axes in meters per second.

To adjust the Kalman filter as a vehicle tracker, the system state vector of the Kalman filter consists of the current vehicle's position (p_x, p_y) and speed (v_x, v_y) within the x/y-plane, respectively.

$$\hat{x}_k = \begin{pmatrix} \hat{p}_x \\ \hat{p}_y \\ \hat{v}_x \\ \hat{v}_y \end{pmatrix} \tag{6.1}$$

To predict position and speed, a vehicle mobility model has to be applied. For the Mobility Data Verification component this model is based on the *equation of linear motion*, whereby, Δt_k denotes the time difference to the previous time step $k-1$. The current position, speed, and acceleration is denoted by p_k, v_k, and a_k, respectively.

$$p_k = p_{k-1} + v_{k-1} \cdot \Delta t_k + a_{k-1} \cdot \frac{\Delta t_k^2}{2} \tag{6.2}$$

As discussed in Section 2.3 CAMs are sent in variable frequencies, i. e., between 1 Hz and 10 Hz. Therefore, Δt_k cannot be assumed to be a constant in this scenario.

According to the introduced equation of linear motion the state transition matrix F_k is represented by a four times four matrix, regarding only position and speed.

$$F_k = \begin{pmatrix} 1 & 0 & \Delta t_k & 0 \\ 0 & 1 & 0 & \Delta t_k \\ 0 & 0 & 1 & 0 \\ 0 & 0 & 0 & 1 \end{pmatrix} \quad (6.3)$$

The acceleration is regarded in neither system state vector nor state transition matrix. However, it is considered as a control factor u_k to the system state before F_k is applied.

$$u_k = \begin{pmatrix} a_x \\ a_y \end{pmatrix} \quad (6.4)$$

Thereby, acceleration in x- and y-direction is calculated from according speed differences of the last received messages and assumed to be constant within each time step. Consequently, it is added with its according factors to the respective speed entries via the control matrix B_k within the Kalman filter prediction.

$$B_k = \begin{pmatrix} 0 & 0 \\ 0 & 0 \\ \frac{\Delta t_k^2}{2} & 0 \\ 0 & \frac{\Delta t_k^2}{2} \end{pmatrix} \quad (6.5)$$

As measurement values for the measurement vector $\tilde{y}_k$ applied to correct the prediction in the correction phase of the Kalman filter, mobility information of received messages are exploited. The position, speed, and heading contained in these messages are converted, analogical as for the system state vector.

$$\tilde{y}_k = \begin{pmatrix} \tilde{p}_x \\ \tilde{p}_y \\ \tilde{v}_x \\ \tilde{v}_y \end{pmatrix} \quad (6.6)$$

Hence, the system state and the measurement vector are similar, such that the measurement matrix H_k is the identity matrix, only. Consequently, Equations 4.4, 4.5, and 4.7 of the Kalman filter can be simplified significantly by eliminating H_k.

$$\Delta y_k = \tilde{y}_k - \hat{x}_k \quad (6.7)$$

$$K_k = P_k \cdot (P_k + R_k)^{-1} \quad (6.8)$$

$$P_k^+ = P_k - K_k \cdot P_k \quad (6.9)$$

The accuracy of the prediction of vehicle mobility data with the applied mobility model heavily depends on current road scenarios. Hence, to calculate the prediction error P_k, the system fault matrix Q_k is chosen dynamically according to the road type, i. e., highway, rural road, or urban road. The measurement variances matrix R_k is analogous chosen according to current GPS accuracy delivered in corresponding messages.

Based on this adaptions and chosen matrices the Kalman filter now can be consulted as a vehicle tracker in the Mobility Data Verification component.

6.2.2. Vehicle Sensor Data Fusion

Under supervision of the author of this thesis, in [QUA11] the Mobility Data Verification component deployed within sim^{TD} field operational trial system architecture is extended by the concept of *vehicle sensor data fusion* as outlined in Section 4.1.2. For this prototypical implementation a radar with a total detection range of about 200 m is advocated.

For sensor data fusion, the position of an object, either indicated by radar or by a C2X message, is transferred into an absolute distance with respect to the host's vehicle position, i. e., the distance from the host vehicle's front to the indicated vehicles rear end. Since sensor and C2X update frequencies are different and not aligned, synchronization of time is considered by means of extrapolation according to the *equation of linear motion*. Thereby, the vehicle's and objects dimensions, as extracted from the C2X messages, are considered. In case of a high update frequency of the own vehicle position and radar measurements, i. e., 10 Hz within sim^{TD}, it is sufficient to exploit object detection tolerance provided by the radar sensor.

Since for all measurements tolerances have to be regarded, at comparing the distance as indicated by a C2X message against the measurement provided by the radar sensor, no exact match is required. In contrast, if the distance indicated by C2X is d_0, the radar sensor's measurement may be within a lower threshold distance d_l and an upper threshold distance d_u to *confirm* the message, as depicted in Figure 6.4. However, if the radar indicates no object or an objects distance above d_u, the message has to be *refused*. On the other hand, if an object is indicated with a distance below d_l, it has to be assumed that a third vehicle is shadowing the investigated vehicle, thus, no decision can be made.

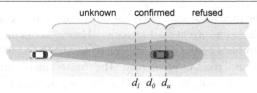

FIGURE 6.4.: The acceptance margin for the Sensor Data Fusion. If the vehicle's
position as indicated by C2X messages is d_0, the radar sensor's
measurement may be within d_l and d_u to confirm the message.

6.3. The *Road Weather Warning* Application in sim^{TD}

In order to evaluate the entire spectrum of capabilities facilitated by C2X communication systems, a multitude of C2X applications are included within the German research project sim^{TD} [SIMTD09b]. As one of the most promising use-cases for C2X communications, the *Weather Hazard Warning* is included within this selection.

As outlined in Section 1.2 the Weather Hazard Warning deploys functionality within vehicles and the infrastructure. Since a comprehensive evaluation of applications is evaluated within sim^{TD}, both parts of the use-case are considered. Accordingly, the application *Road Weather Warning* covers the vehicle-based functionality, whereas the application *Identification of Road Weather* covers the infrastructure-based parts of the Weather Hazard Warning [JH11].

On the one hand, the supporting application Identification of Road Weather[57] is integrated into the framework of the sim^{TD} Central ITS Stations, the *Test Management Center*. As sim^{TD} adaption of the infrastructure-based part of the Weather Hazard Warning application, it focuses on including weather situations, detected by weather measurement stations, into the field operational trial. As outlined in Section 6.1.1, detected weather situations are delivered by the *Straßenzustands- und Wetterinformationssystem* (SWIS). Thereby, six stations in and around the field operational trial area, as listed in Table 6.1 and depicted in Figure 7.9, are consulted. Additionally, all situations detected by vehicles are forwarded to the CIS application. Hence, all detected situations are merged within the application into an overall large scale view on the weather conditions in the field operational trial area.

On the other hand, the application Road Weather Warning[58] is implemented by the author of this thesis as an individual Java/OSGi-bundle and integrated into the

---

[57] sim^{TD} application number: F_1.1.3
[58] sim^{TD} application number: F_2.1.3

OSGi-framework deployed on the sim^TD Application Unit. The application is based upon the advanced situation detection algorithm as presented in Section 3.2.2 and applies a driver notification scheme based upon Chapter 5. A detailed description of the application is available in the respective project documentation in [SIMTD09e] and [SIMTD12a]. However, within this section an overview of the implemented application with respect to previous mentioned concepts is provided.[59]

The Road Weather Warning application is composed out of three of sub-components,

- the Message Listener,
- the Situation Detection including Vehicle Data Listener, and
- the Weather Events,

which are detailed in the following. Additionally, the Road Weather Warning application is part of the *Local Danger Warning* application set[60] [SIMTD09e] and exploits the common *Event Manager* component. This common component triggers maintaining of *Weather Events* containing received or detected weather situations, as detailed in Section 4.5.

As other applications deployed on the AU, the Road Weather Warning application obtains messages via the Relevance Filter system component from the Environment Table. These messages are provided to the *Message Listener* sub-component, which extracts essential data and provides achieved weather situations to the Event Manager.

Within sim^TD, weather situations regarding weather types as listed in Table 6.2 are considered. The cause codes and sub-cause codes are roughly according to TPEC-TEC as specified by [TPEG06] and listed in Table 6.2.[61] Hence, the Relevance Filter is configured to provide only DENMs regarding according situations. All other DENMs, CAMs, and further messages are discarded.

In order to detect weather situations within vehicles, position and local sensor data are consulted within the *Situation Detection* sub-component. Hosts vehicle's position and local sensor data are provided by the *VAPI* system component. Within

---

[59] Within the context of this thesis, the terms *situation* and *event* denote single individual detections and merged groups of situations, respectively, as introduced in Chapter 3. However, within sim^TD a situations is denoted as *Event* and an event is denoted as *Aggregat*.

[60] sim^TD application set number: HF_2.1

[61] For *Heavy Rains* situations, the sub-cause *unknown* is exploited, since no distinction between *heavy rain* (code 45 / 1), *heavy snowfall* (code 45 / 2), or *soft hail* (code 45 / 3) is made. Accordingly, for *Icy Road* situations, the sub-cause *unknown* is exploited, too, since no distinction between the actual reason, either *heavy frost* (code 11 / 1), *snow* (code 11 / 4), *ice* (code 11 / 5), or *black ice* (code 11 / 6), is made.

Situation Type	Cause / Sub-cause	Code
Heavy Rains	precipitation / unknown	45 / 0
Dense Fog	visibility reduced / due to fog	43 / 1
Icy Road	slippery road / unknown	11 / 0
Cross Winds	extreme weather / strong winds	42 / 1

TABLE 6.2.: Considered weather types in sim^{TD} and their respective TPEG-TEC cause code and sub-cause code.

the Situation Detection the *Vehicle Data Listener* module receives the data and provides them to the detection algorithm. In contrast, the local time is achieved directly from the operating system wherever needed. This system time is adjusted via NTP to the GPS time as detailed in Section 6.1.2 and highly synchronized between different ITS Stations.

However, the advanced situation detection algorithm as detailed in Section 3.2.2 is slightly adapted and simplified, due to the available vehicle sensor data and regarded weather types. Hence, no detection of *Aquaplaning* is implemented and, since no distinction between precipitation of rain or snow is made, detection of snowfall is omitted. Finally, since the distinction between TCS, ABS, or ESP activation is not available within all vehicles, at slippery road detection all three driving maneuvers, i. e., acceleration, deceleration, and during driving, are merged and any requirement on pedals is omitted. However, optional conditions and information quality levels are applied, if available.

The Situation Detection implements the detection algorithm with according trigger conditions for each of the three resulting weather types in respective individual modules, i. e., Rain Detection, Fog Detection, and Ice Detection. Detected situations are provided to the Event Manager component, where they are integrated into an according Weather Event.

In order to simplify internal calculations, incoming WGS 84 coordinates, either within messages or host vehicle's position, are converted by respective subcomponents, i. e., the Message Listener or the Vehicle Data Listener, into a two dimensional Cartesian coordinate system, as detailed in Section 2.5.2. Since within sim^{TD} computations resources are not hard limited, the comparatively complex UTM-coordinates are exploited. Additionally, in the area of the sim^{TD} field operational trial, there is no UTM zone border to be considered. Thus, no handling of coordinates located in different zones is required.

Once the common component *Event Manager* receives a weather situation, either from the Message Listener or the Situation Detection, all *Weather Events* are polled

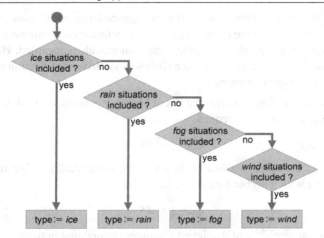

FIGURE 6.5.: Hierarchy of the weather types selected for a Weather Event if weather situations with different types are included.

to verify if the situations shall be included, as detailed in Section 4.5.

Since weather situations of different weather types are detected independently from each other, actual weather conditions may apply to multiple triggering conditions at the same time. Hence, situations with different weather types may exist for the same area and, thus, within one Weather Event. However, the weather type of the resulting event is applied according to the rule as depicted in Figure 6.5. On the one hand, this hierarchy regards that some weather types may most probably imply another weather type, e. g., rainy weather conditions are also implying reduced visibility and strong winds. On the other hand, the underlying possible danger is considered, e. g., an icy road may have a more severe impact on road safety than winds.

The Weather Event sub-components maintain included situations. Therefore, they are triggered by the Event Manager periodically.

Hence, during such an update cycle, if a new detected situation is included in the event, respective information is forwarded to the Communication Client system component. Thus, a C2X Message object is created and a corresponding DENM is sent. Since forwarding and repeating of messages is maintained by the C2X Facilities, this is required only once for detected situations and never for received situations.

As a last step during the update cycle, it is verified if the vehicle is approaching the Weather Event's area. According to the driver notification strategies detailed in

Chapter 5, either a new driver notification is triggered, if a notification is already present, it is updated, or, if the event is reached, the notification is removed. Thereby, the current state of the respective weather detection module is regarded. Hence, if a detection module is currently detecting a situation, no driver notification regarding the same weather type is triggered.

Finally, if no presentation is active, the Event Manager is notified to call the event's update cycle again with low frequency, i. e., after

$$t_{upd_{low}} = 5 \, \text{s} \quad . \tag{6.10}$$

Otherwise, i. e., if a driver notification is active, the event shall be triggered by the Event Manager with a higher frequency, i. e., after

$$t_{upd_{high}} = 2 \, \text{s} \quad , \tag{6.11}$$

thus, notification levels[62] and displayed distances are updated in time.

A simplified overview of the Road Weather Warning application as well as its major components and interconnections is depicted in Figure 6.6.[63]

Driver notifications are displayed on the sim^{TD} HMI, i. e., a 7" automotive display with audio and touch-screen, mounted into vehicles. The layout of the presentation is developed by *Deutsches Forschungszentrum für Künstliche Intelligenz* [DFKI] with respect to instructions of the author of this thesis. As detailed in [SIMTD09d], the layout consists of three selectable screens:[64]

1. the *navigation screen*, displaying a map, the test route, and driving instructions,

2. the *main screen* for displaying driver notifications, and

3. the *report screen*, facilitating the driver to report objects on the street.

Regardless of the current selected screen, if a safety related driver notification is requested, the HMI switches to the main screen.

Only one notification, i. e., the notification with the highest priority, is displayed at the main screen at a time. However, the main screen consists of a *icon bar* located on the top. Within this bar, an icon for each active notification is displayed and might be selected by the driver.

---

[62] Within sim^{TD}, the lowest notification, which does not trigger a presentation on the main screen, is denoted with *Possible*, instead of *Background*.

[63] Additional management, configuration, development and debugging, or field operational test related components are omitted.

[64] In fact, there is an additional *debug screen*. However, this screen is intended for system development, only, and not part of the HMI layout

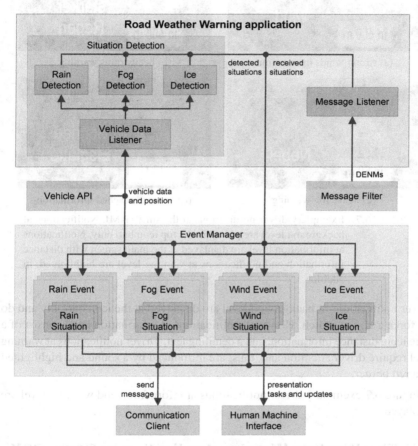

FIGURE 6.6.: Architecture of the *Road Weather Warning* application, with respect to implemented essential components as well as their internal and external interconnections.

(a) strong winds information

(b) heavy rains warning

(c) dense fog warning

(d) icy road reached information

FIGURE 6.7.: Exemplary driver notifications at the sim^{TD} HMI. Notifications at background level are placed in the top icon bar, only. Notifications at information level are displayed at the main screen with distance information and notifications at warning level are highlighted, in addition.

Driver notifications at background level are displayed at the icon bar only and do not force a screen switch. However, notifications at information level consist of a large icon, distance information, and a detailing text. Driver notifications at warning level require driver attention and, thus, are introduced by a sound and highlighted by a red border.

In Figure 6.7 exemplary driver notifications at information and warning level are displayed.

6.4. The *Weather Warning* Application in DRIVE C2X

Within the European research project DRIVE C2X, leading car manufactures and automotive suppliers join together to build up a field operational trial on multiple locations [DRIVE11b] spread all over Europe. Thereby, no dedicated Vehicle ITS Station or Roadside ITS Station is mandatory. Instead, multiple diverse VIS and RIS architectures are deployed. Hence, the interoperability of different C2X communication systems developed so far is evaluated.

However, a reference implementation providing a VIS system with basic system components [DRIVE11a], e. g.,

- a *Local Dynamic Map* (LDM), the message container,
- a *Communication Client*, sending messages,
- a *Vehicle Data Provider* (VDP), providing vehicle sensor data,
- a *Position and Time* (PoTi), providing GPS positioning and time, and
- a *Human Machine Interface* (HMI), facilitating presentations to the driver,

is jointly developed together with reference implementations of a basic set of C2X applications [DRIVE12b]. As a promising use-case, the Weather Hazard Warning is included into this set of selected applications. Hence, the DRIVE C2X *Weather Warning* application is implemented and included into the DRIVE C2X system by the author of this thesis.

Compared to the sim^{TD} adaption of the Weather Hazard Warning application, as presented in Section 6.3, the application architecture of the DRIVE C2X adaption is adjusted to the respective system architecture and enhanced and improved. This section provides an overview of the basic application concepts and architecture, whereas a more detailed description is available at [DRIVE11c] and [DRIVE12c].

In contrast to [ETSI10a] or [C2C07] the generic DRIVE C2X Vehicle ITS Stations reference architecture intends to integrate all applications and system components into one physical unit. This *Application Unit* may be connected directly to, e. g., HMI, ITS-G5A antenna(s), GPS receiver, or vehicle CAN(s), without the need of a CCU. This promising approach is adapted in different degrees by the various VIS system architectures within DRIVE C2X [DRIVE12a].

The DRIVE C2X reference system, exploits an OSGi-framework hosting all applications and system components. Hence, the *Weather Warning* application is implemented as an individual Java/OSGi-bundle relying on services provided by the various system components. Thereby, the application architecture is separated into multiple internal sub-components, implementing diverse required tasks.

Accordingly, incoming C2X messages are provided by the *Local Dynamic Map* (LDM) system component. Since there is no relevance filter, all DENMs received by the VIS are provided to the application. Consequently, message filtering is done by the *Message Listener* sub-component. Thereby, weather situations as listed in Table 6.3 are considered. Exploited DENM cause codes and sub-cause codes are according to [ISO13]. Upon reception, regarded weather situations are forwarded to the application's *Event Manager* sub-component.

Situation Type	Cause / Sub-cause	Code
Dense Fog	visibility reduced / due to fog	18 / 1
Heavy Snowfall	precipitation / heavy snowfall	19 / 2
Heavy Rains	precipitation / heavy rain	19 / 1
Aquaplaning	aquaplaning / no sub-cause	7 / 0
Icy Road	slippery road / heavy frost	6 / 1
Cross Winds	extreme weather / strong winds	17 / 1

TABLE 6.3.: Considered weather types in DRIVE C2X and their respective TPEG-TEC cause code and sub-cause code.

Vehicle sensor data are provided by the *Vehicle Data Provider* (VDP) system component, whereas the current vehicles position and synchronized time are provided by a *Position and Time* (PoTi) system component. However, within the Weather Warning application the *Vehicle Data Listener* sub-component receives data from both system components and provides them to other application sub-components. Hence, the separation of VDP and PoTi is hidden and all host vehicle related data are internal provided by one sub-component.

In order to simplify internal calculations, incoming WGS 84 coordinates, either within messages or host vehicle's position, are converted by respective sub-components, i. e., the Message Listener or the Vehicle Data Listener, into a two dimensional Cartesian coordinate system, as detailed in Section 2.5.2. Within DRIVE C2X the DLE coordinates, as proposed by the author of this thesis, are exploited for internal coordinate representation. Hence, fast and highly efficient coordinate transformation with high accuracy due to low distortion is facilitated.

The Weather Warning application implements the advanced detection algorithm as presented in Section 3.2.2. Thereby, respective trigger conditions are integrated into one *Situation Detection* sub-component, relying on and considering all available vehicle data. Thus, the most appropriate weather type for the current weather condition is determined.

Once a weather situation is detected by the Situation Detection, relevant information is provided, on the one hand, immediately to the *Message Sender* sub-component. This information includes trace points, indicating the path the vehicle has passed during detection. The Message Sender subsequently builds up an according C2X message and requests the sending of a respective DENM via the *Communication Client* system component.

On the other hand, detected situations are provided to the application's *Event Manager* sub-component. Thus, all available weather situations, either from the

Message Listener or the Situation Detection, are collected and maintained by the Event Manager. Accordingly, within the Event Manager, situations are integrated into known *Weather Events* if their areas do overlap, as detailed in Chapter 4, or into a new **Weather Event**. Thereby, the included trace is consulted as detailed in Section 4.3.4.

Within *Weather Events*, it is periodically evaluated if the host vehicle is approaching the border of an event. If not, the evaluation is done again with low frequency, i. e., after

$$t_{upd_{low}} = 2\,\text{s} \quad . \tag{6.12}$$

Otherwise, i. e., if the host vehicle is approaching the border of an event, a driver notification is requested or, if already active, updated accordingly with a higher frequency, i. e., after

$$t_{upd_{high}} = 1\,\text{s} \quad , \tag{6.13}$$

thus, notification levels and displayed distances are updated in time. Thereby, the *Display Control* sub-component is exploited as interface to the *Human Machine Interface* system component.

A simplified overview of the major components and interconnections, both, internal and to the DRIVE C2X generic reference system architecture, of the Weather Warning application is depicted Figure 6.8.[65]

In addition to the presented Weather Warning application, within DRIVE C2X a *Weather Warning RIS* application is implemented by the author of this thesis. This RIS application is exploited as interface for infrastructure-based weather situation detection. Thus, both, the vehicle-based as well as the infrastructure-based parts of the Weather Hazard Warning are included within DRIVE C2X.

Thereby, once a weather situation is detected by infrastructure components, e. g., via weather measurement stations or a weather service provider, according information, i. e., weather type, position, validity duration, and affected area, may be provided to the Weather Warning RIS application via a respective *API*. Subsequently, corresponding *Weather Situations* are maintained at the *Distributor* sub-component. It maintains situation updates and revocations triggered by the infrastructure and requests sending of according C2X messages at the *Communication Client* system component via the *Message Sender* sub-component.

A simplified overview of the major components and interconnections, both, internal and to the DRIVE C2X RIS reference system architecture, of the Weather Warning

---

[65] Additional management, configuration, development and debugging, or field operational test related components are omitted.

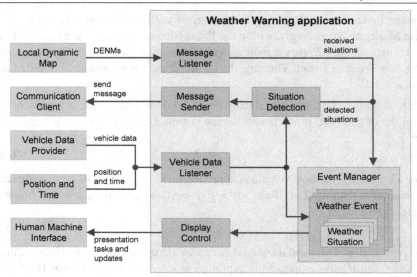

FIGURE 6.8.: Architecture of the *Weather Warning* application, with respect to implemented essential components as well as their internal and external interconnections.

RIS application is depicted Figure 6.9[66]

Due to the multitude of different C2X systems integrated into DRIVE C2X, the hardware and layout of the HMI and the presentations may vary between different test sides [DRIVE13a] and car manufactures participating the project. However, a reference HMI is provided by the *Fraunhofer-Institut für Offene Kommunikationssysteme* [FOKUS] regarding inputs of the author of this thesis. This reference HMI presents driver notifications on a tablet computer with 7" touch-screen display mounted into the vehicle. The HMI-Application executed within the *Android* system of the mobile device is connected via *WiFi* to the respective *HMI Service* OSGi bundle within the Application Unit. The default main screen consists of either a map, indicating the position of the vehicle and the test route, or a test message with instructions stated by the test management.

Driver notifications at background level are integrated as small icons at the respective position within the map screen, only. For weather related situations, the icon is positioned at the center of the situation's area. Notifications at information

---

[66] Additional management, configuration, development and debugging, or field operational test related components are omitted.

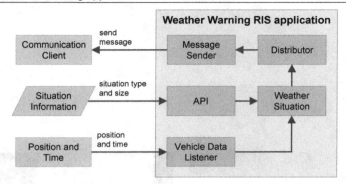

FIGURE 6.9.: Architecture of the *Weather Warning Roadside ITS Station* application, with respect to implemented essential components as well as their internal and external interconnections.

level are presented as larger icon in the upper right corner of the main screen, as exemplary depicted for a strong winds information in Figure 6.10 (a). If multiple driver notifications at information level are requested, icons are stacked on top of each other. However, if a driver notification is requested at warning level, a large icon with additional information, e. g., the distance, is displayed on the right hand side of the screen. Thereby, no notifications at information level are displayed anymore, as depicted in Figure 6.10. Since a notification at warning level requires a certain level of driver attention, it is introduced by a sound and only one notification, i. e., the notification with the highest priority, is displayed at a time. Finally, for notifications at severe level, no map or text message is displayed furthermore. In contrast, only notification related content is visible. [67]

---

[67] However, as detailed in Section 5.2.1, driver notifications at severe level are not exploited by the Weather Hazard Warning application.

(a) strong winds information with map

(b) aquaplaning warning with map

(c) heavy snowfall warning with text

(d) heavy rains warning with map

FIGURE 6.10.: Exemplary driver notifications at the DRIVE C2X reference HMI.
The background consists of either a map or text. Driver notifications
at information level are displayed as icon in the upper right corner,
notifications at warning level are illustrated with a large icon and
distance information.

7. Evaluation Results

In this chapter approaches, modules, and components as detailed in previous chapters are evaluated. Therefore, measurement data obtained within driving simulation, individual tests with especially equipped and adapted vehicles, and large scale field operational trials are consulted.

7.1. DLE Coordinates Applicability

As detailed in Section 2.5.2, for internal calculations received WGS 84 coordinates are converted into a two dimensional Cartesian coordinate system. Therefore, the UTM coordinate system as presented in Section 2.5.2 provides a feasible solution. Thus, this approach is implemented in the *Road Weather Warning* application within the large scale field operational trial sim^{TD}, as detailed in Section 6.3. However, due to the elaborate algorithm to achieve UTM coordinates, as described in [DMA89], expensive calculations are required. Thus, a huge amount of computational resources is consumed.

Consequently, the author of this thesis proposes DLE coordinates as introduced in Section 2.5.2, which can be highly efficiently calculated by exploiting the distance to a reference meridian λ_0 as the position's x-coordinate and the distance to the Equator as its y-coordinate. Thereby, the earth's surface is assumed as a perfect sphere with radius r of

$$r = 6\,371\,000.8\,\text{m} \quad . \tag{7.1}$$

This approach is implemented in the *Weather Warning* application within the European research project DRIVE C2X, as detailed in Section 6.4.

As during every coordinate transformation from geographical coordinates to a plane two dimensional coordinate system, distortion affecting the accuracy has to be regarded. However, Car-to-X communication considers events within the near environment around the host vehicle. Thus, conversion accuracy has to be within acceptable margins nearby only. In this section the absolute inaccuracy of

$$F_1 = \frac{\varphi_1 + \varphi_2}{2} \quad , \quad F_2 = \frac{\varphi_1 - \varphi_2}{2} \quad , \quad L = \frac{\lambda_1 - \lambda_2}{2}$$

$$S_1 = \cos^2 F_1 \cdot \sin^2 L + \sin^2 F_2 \cdot \cos^2 L$$
$$S_2 = \sin^2 F_1 \cdot \sin^2 L + \cos^2 F_2 \cdot \cos^2 L$$

$$\omega = \arctan\sqrt{\frac{S_1}{S_2}} \quad , \quad \rho = \frac{\sqrt{S_1 \cdot S_2}}{\omega}$$

$$d_0 = 2 \cdot a \cdot \omega$$

$$H_1 = f \cdot \sin^2 F_1 \cdot \cos^2 F_2 \cdot \frac{3 \cdot \rho - 1}{2 \cdot S_2}$$

$$H_2 = f \cdot \cos^2 F_1 \cdot \sin^2 F_2 \cdot \frac{2 \cdot \rho + 1}{2 \cdot S_1}$$

$$d = d_0 \cdot (1 + H_1 - H_2)$$

cf. [MEE98]

FIGURE 7.1.: Algorithm to determine the orthodromic alike distance between two geographical positions with latitudes φ_i and longitudes λ_i on an ellipsoid. Thereby, the ellipsoid's major radius a, i. e., the earth's equatorial radius, and flattening f is regarded.

the proposed DLE coordinates and the relative error compared to UTM coordinates are evaluated.

As the metric to assess the accuracy of the proposed DLE coordinates, the error Δ_{err} in calculating the distance between two geographical positions is consulted. Hence, for any two positions, the distance calculated within DLE coordinates is compared to a most highly accurate reference distance d.

For two points P_1 and P_2 located on the surface of a sphere, the distance measured along the surface of the sphere is the orthodromic distance, which is along the great-circle defined by these two points. However, since the earth is more appropriate approximated by an ellipsoid, the reference distance d is determined by an orthodromic alike distance, i. e., the shortest distance between the positions on the surface of the WGS 84 ellipsoid. The best known solution [MEE98] to determine the orthodromic distance on an ellipsoid is detailed in Figure 7.1, which is based on [BDL50]. Thereby, the ellipsoid's major radius, i. e., the radius at the equator, and flattening is regarded. According to the WGS 84 reference ellipsoid, the earth's major radius a and the earth's flattening f are

$$a = 6378\,137\,\text{m} \tag{7.2}$$

$$f = \frac{1}{298.257\,222\,563} \quad , \tag{7.3}$$

respectively. Accordingly the reference distance d is assumed to be the real distance between two geographical positions P_1 and P_2 with latitudes φ_1 and φ_2 as well as longitudes λ_1 and λ_2.

In order to apply the detailed metric, the calculated distance d_{DLE} between the two positions by exploiting DLE coordinates is determined in addition. Thereby, the respective WGS 84 coordinates are converted into DLE coordinates, such that P_1 and P_2 are represented by respective Cartesian x- and y-coordinates (x_1, y_1) and (x_2, y_2). Thus, the distance may be determined by applying the Pythagorean theorem

$$d_{DLE} = \sqrt{(x_2 - x_1)^2 + (y_2 - y_1)^2} \quad . \tag{7.4}$$

Consequently, the distance error Δ_{err} is given by

$$\Delta_{err} = d - d_{DLE} \quad . \tag{7.5}$$

For DLE coordinates, the distortion varies according to the relative direction. Accordingly, the distance error is minimal, if both points are located on the same meridian, i. e., sharing the same longitude and differ in north-south direction only. In contrast, the error is maximal, if they are located orthogonal to the meridians, i. e., sharing the same latitude and differ in east-west direction only.

However, in Figure 7.2 the mean distance error $\overline{\Delta}_{err}$ and the error variation is displayed at a logarithmic scale for increasing distance. All consulted positions are located in the area around Frankfurt am Main, Germany, at a latitude $\varphi \approx 50.1°$ and longitude $\lambda \approx 8.7°$.

Accordingly, the relative distance error $\Delta_{\widetilde{err}}$ does not exceed $0.32\,\%$, which are $3.2\,\text{m}$ at a distance of $1000\,\text{m}$. Thereby, the mean distance error is at $0.20\,\%$, or $2.0\,\text{m}$ at a distance d of $1000\,\text{m}$. Moreover, even at a larger scale, the distance error Δ_{err} grows approximately linear with the increasing distance, as depicted in Figure 7.3, such that the relative distance error stays constant within up to $10\,\text{km}$.

In contrast, the evaluation with positions located around the globe shows that the relative distance error is not equal at any latitude. Instead, the relative error $\Delta_{\widetilde{err}}$ varies from up to $-0.4\,\%$ for positions located near the Equator and $\approx 0.45\,\%$ for positions near the poles, as depicted in Figure 7.4. At a distance of $1000\,\text{m}$, this error

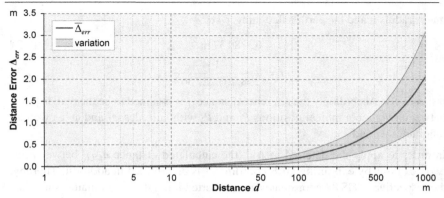

FIGURE 7.2.: Mean distance error $\overline{\Delta}_{err}$ and error variation for DLE coordinates at
a logarithmic scale for medium ranges up to 1000 m.

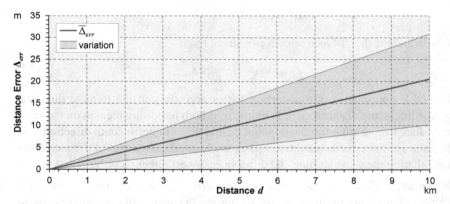

FIGURE 7.3.: Linear increasing mean distance error $\overline{\Delta}_{err}$ and error variation for
DLE coordinates for large distances up to 10 km.

is not more than -4 m with a mean of -1.2 m at the Equator and approximately
4.5 m at the poles.

Finally, the accuracy of the proposed DLE coordinates is compared to the accuracy
of the well-known UTM coordinate system. Thereby, the respective WGS 84
coordinates are converted into the UTM coordinate system, such that P_1 and P_2 are
represented by respective easting- and northing-components. These are exploited as
Cartesian x- and y-coordinates (x_1, y_1) and (x_2, y_2), respectively. Thus, the distance
between the two positions may be again determined by applying the Pythagorean

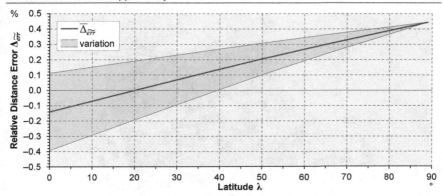

FIGURE 7.4.: Mean relative distance error $\Delta_{\widetilde{err}}$ and error variation for DLE coordinates along the globe.

theorem

$$d_{UTM} = \sqrt{(x_2 - x_1)^2 + (y_2 - y_1)^2} \quad . \tag{7.6}$$

Accordingly, the difference Δ_{UTM} in the calculated distance by exploiting DLE
coordinates or UTM coordinates, is given by

$$\Delta_{UTM} = d_{DLE} - d_{UTM} \quad . \tag{7.7}$$

Since UTM coordinates suffer by distortion as DLE coordinates, this difference is
not an absolute distance error as Δ_{err}, but a relative error between the two coordinate
systems.

However, distortion for UTM coordinates heavily varies with the actual position
within the respective UTM zone. Hence, Δ_{UTM} is exemplary calculated for diagonal
distances, i. e., same distance in latitudinal and longitudinal direction, of 1000 m
within the UTM zone 32 U, which covers large areas of Germany. As depicted
in Figure 7.5, the difference between the distance calculated exploiting UTM
coordinates and calculated exploiting DLE coordinates does not exceed 1.5 m, with
a mean difference of 0.8 m, at a real distance of 1000 m.

This evaluation of the DLE coordinates, with a metric relying on the distance calculation between geographical positions, demonstrates the accuracy of the proposed
approach. For medium ranges, reliable calculations can be done by exploiting DLE
coordinates with negligible low inaccuracy. Especially, compared to the dispro-
portional more complex task to calculate UTM coordinates, the DLE coordinates
provide a computation cost effective and fast to calculate alternative solution.

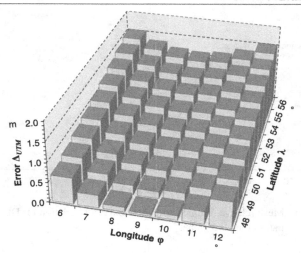

FIGURE 7.5.: The relative difference Δ_{UTM} between the calculated distance by exploiting DLE and UTM coordinates for distances of $d \approx 1000$ m. Regarded positions are located in the UTM zone 32 U.

7.2. Mobility Data Verification Accuracy

The Mobility Data Verification component as introduced in Section 4.1 is implemented as individual Java/OSGi bundle and integrated into the field operational test platform of the German research project sim[TD] as detailed in Section 6.2. To prove functional correctness and performance, tests and evaluations by means of different test drives and multiple recorded real word traces were carried out.

Accordingly, test drives in all relevant road classes, i.e., urban, rural, and highway scenarios, are performed and evaluated. In Figure 7.6 just one exemplary route, which covers several different road classes in one run, is depicted.

In order to determine and compare prediction accuracies at different CAM frequencies, real world traces are recorded at 10 Hz frequency during such test drives, in a first step. Subsequently, the component is supplied with messages at different frequencies generated out of these recorded traces by applying the corresponding CAM generation algorithm to them. Hence, realistic real world messages at static frequencies, i.e., 1 Hz, 2 Hz, 10 Hz, and the dynamic frequency according to ETSI [ETSI11b] are carried out.

As depicted in Figure 7.7, the replaying results show that the prediction accuracy

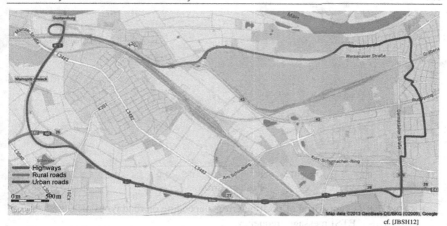

cf. [JBSH12]

FIGURE 7.6.: Exemplary route for one test drive, covering highways, rural roads, and urban areas in and near Rüsselsheim, Germany.

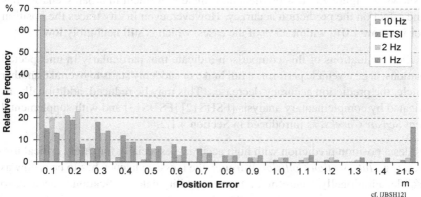

cf. [JBSH12]

FIGURE 7.7.: Comparison of measured position errors depending on different message frequencies.

of the advocated Kalman filter-based vehicle tracker is best at the highest CAM frequency of 10 Hz. Thereby, the prediction error is below 0.3 m within 95 % of the cases. With variable CAM intervals according to ETSI, a prediction error of 0.9 m in 95 % of the cases is achieved. This result is significantly lower than the GPS accuracy.

Additionally, the effect of different road classes on the prediction error is evaluated. Therefore, multiple test drives on highways and cities are compared, each performed with the dynamic CAM frequency.

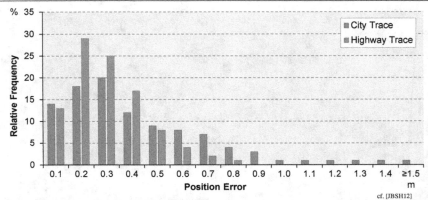

cf. [JBSH12]

FIGURE 7.8.: Comparison of position errors depending on road types measured at
ETSI message frequency.

As shown in Figure 7.8, less predictable vehicle movement in urban scenarios has some effect on the prediction accuracy. However, even in city traces the position error is below 1.0 m within 95 % of the cases, which is still negligibly low.

Detailed evaluations of this comparison indicate that particularly in unexpected situations, e. g., a vehicle performs a full brake or suddenly starts to overtake another vehicle, the prediction accuracy decreases. This may be reduced, additionally, as indicated by complementary analysis [FSHS12] [FSHS13] and with supplemental *Local Sensor Checks*, as introduced in Section 4.1.2.

Besides a position prediction with high accuracy, especially for safety critical use-cases, e. g., *Intersection Collision Warning*, the message latency has to be as low as possible, additionally. Therefore, efficient mobility data verification is needed, so that messages are not delayed more than necessary.

During the multiple examined tests, as described above, an average overall message latency of about 2.7 ms was achieved on the sim[TD] field operational test hardware. Thereby, function call, quick checks, and evaluation takes about 1.0 ms, whereas Kalman filter prediction and correction take the remaining 1.7 ms.

Finally, the implementation of the Mobility Data Verification component as detailed in Section 6.2 is integrated into the German research project sim[TD]. This way, the component contributes to ensure trustworthy messages during the field operational trial phase of the project and, hence, may serve to improve reliability of application evaluation results.

7.3. Situation Detection Reliability

As detailed in Section 6.3, an adaption of the proposed Weather Hazards Warning application, the *Road Weather Warning*, is part of the long term large scale field operational trial of the national research project sim^{TD}. Thus, for a duration of about six months, the application, developed and implemented by the author of this thesis, operates within various test runs by up to 120 vehicles. Within this chapter, results from the field operational tests are consulted to evaluate the suitability of the weather event detection.

Since weather events are hard to forecast and, hence, cannot be included into a large scale field operational trial schedule, only few planned weather related test are examined [SIMTD10d]. In contrast, as detailed in [SIMTD09l] and [SIMTD10c], evaluation of weather detection is focused on randomly arising weather events. Hence, the Weather Hazard Warning is running during almost all sim^{TD} test runs in addition to the investigated C2X use-case. Thus, a realistic event detection behavior of the application based on real world weather situations is achieved.

Within sim^{TD}, field operational test trials are examined on highways, rural roads, and urban roads distributed all over the Rhein-Main metropolitan area in and around Frankfurt am Main, Germany. This field operational test area, including the positions of the consulted weather measurement stations, is depicted in Figure 7.9.

Situation Detection Correctness

As outlined previously, the sim^{TD} adaption of the Weather Hazard Warning application is operating while vehicles are driving in the field operational test area. To facilitate later evaluation, during execution of field operational trials relevant system and application states are recorded. Hence, once a rain, fog, or slippery road weather situation arises, the application's algorithms deployed within test vehicles start to detect situations, process data, send DENMs, and generate driver notifications.

The later evaluation consults reports provided by test drivers, rain radar images, and comparison between vehicles operating at the same time in the same area. As detailed in [SIMTD13c] and [SIMTD13f], it is based on about

- 2 000 individual dense fog situations,
- 15 000 individual slippery road situations, and
- 65 000 individual heavy rain situations

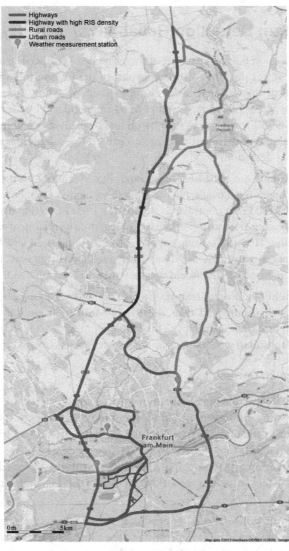

cf. [SIMTD10d]

FIGURE 7.9.: Field operational test area of the German research project sim^{TD}, which includes highways, rural roads, and urban regions in the Rhein-Main metropolitan area in and around Frankfurt am Main, Germany.

detected by vehicles.[68]

Regarding the ratio of not detected situations, i. e., false-negatives, the reference, i. e., driver reports and rain radar images, indicate a negligible low rate for all three situation types. Hence, if there is a weather situation indicated by the reference, the application detects it with high probability. However, it has to be regarded that the major reference for dense fog and slippery road situations, i. e., test driver reports, is not perfectly reliable. For heavy rain situations, a comparable reliable reference, i. e., rain radar images, is consulted.

Regarding the ratio of detected situations, where no hazardous weather situation is present, i. e., false-positives, the result depends on the situation type. For dense fog situations, only about one third of the detected situations can be verified to be surely correct. However, within sim^{TD} the fog detection mainly depends on the activation of the rear fog light. As the evaluation shows, once a rear fog light is activated, drivers tend to leave it unintentionally activated. This results in the low ratio as further detailed in [SIMTD13f]. Accordingly, for future implementations additional sensors have to be regarded for dense fog detection.

On the other hand, for slippery road and heavy rain situations, test driver reports, correlation between vehicles, and rain radar images indicate a negligible low rate of false-positive detections. However, one major reference, the test driver reports, is not perfectly reliable, again. But, since humans do notice false-positives relatively reliable, it is suitable for this case.

Situation Area Sharpness

Within the Weather Hazard Warning application, rectangles are exploited to encode a situation's area, as detailed in Section 3.2.3. In the following, the accuracy of rectangles covering the shape of a road is evaluated. Therefore, a metric based on the covered area is consulted. Hence, for an exemplary real world scenario the area of a highway is compared with the area of the situation's rectangles. Additionally, two circle-based approaches are evaluated against the metric and the rectangular-based approach.

Accordingly, within the selected scenario, i. e., a field operational test run along the highway A5 near Friedberg, Germany, a rain event is detected by a vehicle driving from north to south. The overall detection length is about 7.1 km. With a mean

---

[68] The infrastructure-based situation detection in sim^{TD} is not implemented by the author of this thesis and, hence, respective evaluation is not further detailed. However, an extensive evaluations of the infrastructure-based situation detection in sim^{TD} is provided by the author of this thesis in [SIMTD13e].

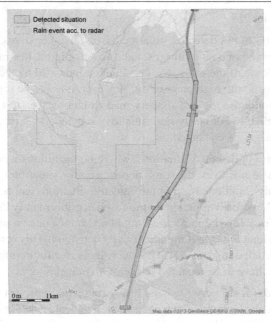

cf. [JH13]

FIGURE 7.10.: Exemplary field operational test run near Friedberg, Germany. The
vehicle is driving from north to south along the highway A5 and
is incrementally detecting a heavy rain situation. Each situation is
represented by a rectangle.

6-line highway width of 36 m [FGSV08], an overall highway area of about

$$A_{ref} = 255\,600\,\text{m}^2 \tag{7.8}$$

is assumed as reference area.

As detailed in Section 3.2.2, the vehicle is incrementally detecting the heavy rain
event. Figure 7.10 depicts the generated rectangles during the exemplary field
operational test run. Additionally, the rain events expansion as reported by a rain
radar is illustrated. The image clearly shows how sharp the highway is covered by
the generated rectangles and how precisely the beginning of the weather event is
encountered.

Regarding the detected situations, the mean situation width $\overline{w}_{rect}$, the mean situation length $\bar{l}_{rect}$, and the mean situation area $\bar{A}_{rect}$ are about

$$\bar{w}_{rect} = \quad 103.3\,\text{m} \quad , \tag{7.9}$$

$$\bar{l}_{rect} = \quad 971.1\,\text{m} \quad , \text{and} \tag{7.10}$$

$$\bar{A}_{rect} = 100\,300\,\text{m}^2 \quad . \tag{7.11}$$

As detailed in Section 3.2.2 and Section 4.3 detected situations overlap. Within the current scenario, the mean overlapping is about 10 % of a rectangle's area, such that the overall covered area is about

$$A_{rect} = 722\,200\,\text{m}^2 \quad . \tag{7.12}$$

Thus, the rectangles cover about 282 % of the reference highway area A_{ref}.

Another common approach, utilizes circles to encode situation areas. Applying this shape to the investigated scenario, circles might have a diameter equivalent to the length of the rectangles, i. e., a mean radius $\bar{r}_{circ_a}$ of about

$$\bar{r}_{circ_a} = \quad 485.5\,\text{m} \quad . \tag{7.13}$$

Consequently, the mean circular situation area is about

$$\bar{A}_{circ_a} = 740\,500\,\text{m}^2 \quad , \tag{7.14}$$

which is already above the overall area of the rectangular-based approach.

However, within this adaption, the overlapping area covers about 3.7 % of the circle's area, such that the overall covered area A_{circ_a} is about

$$A_{circ_a} = 5\,703\,000\,\text{m}^2 \quad . \tag{7.15}$$

Thus, the circles cover about 2230 % of the reference highway area, i. e., roughly ten times as much as the rectangle-based solution.

In Figure 7.11(a) this alternative interpretation of the exemplary field operational test run is depicted.

In addition to the huge amount of unintended covered area, the width of the area varies heavily. To mitigate this effect, generated situations might overlap more. On the other hand, to reduce the covered area, the frequency at which situations are generated might be increased.

Consequently, for a second alternative interpretation, a doubled situation generation frequency with the situation's radius reduced to 2/3, i. e., a mean radius $\bar{r}_{circ_b}$ of about

$$\bar{r}_{circ_b} = \quad 323.7\,\text{m} \quad , \tag{7.16}$$

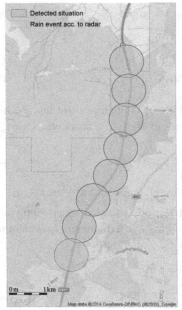
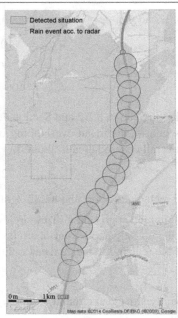

(a) Circles, instead of rectangles   (b) Circles, more frequent updates

FIGURE 7.11.: Alternative interpretations of the situations depicted in Figure 7.10. Within (a) each situation is represented by a circle with same diameter as the rectangle's length. Within (b) the situation generation frequency and situation overlapping are increased such that the tube is thinner and smoother.

is assumed. Thus, the mean situation area $\overline{A}_{circ_b}$ is about

$$\overline{A}_{circ_b} = 329\,100\,\text{m}^2 \quad . \tag{7.17}$$

Concerning the increased situation overlap, the overlapping area is about 20 %. Accordingly, the circles of this second alternative cover an overall area A_{circ_b} of about

$$A_{circ_b} = 4\,205\,000\,\text{m}^2 \quad , \tag{7.18}$$

which is about 1645 % of the reference highway area.

As depicted in Figure 7.11(b), this second alternative builds up a significant smoother and smaller tube. However, the additional covered area is still considerably larger, i.e., more than five times as large, as with the rectangular solution. In addition, it has to be considered that the increased situation generation frequency has a substantially impact on the message load within the communication channel. Consequently, by exploiting circles as shapes for situations significant area beside the road is covered or the number of messages has to be increased. Especially, if higher *ASIL* levels [ISO11] have to be fulfilled, high location accuracy and low channel occupation is required. Accordingly, even if the first impression on exploiting rectangles might be unfamiliar, this shape provides important advantages on both, information accuracy and channel load.

7.4. Driver Notification

As detailed previously, the major goal of the Weather Hazard Warning application as presented in this thesis is to notify a vehicle's driver about possible dangerous upcoming weather related road conditions substantially influencing the road safety. Thus, the driver might adapt the driving behavior accordingly as further detailed in Chapter 5. However, to facilitate such driver information, a notification at the right point in time is mandatory.

To evaluate, if the *recognition field* as detailed in Section 5.1.2 is suitable to facilitate driver notifications on time, real world scenarios have to be considered. Hence, results of the field operational trial of the European research project DRIVE C2X are investigated.

In order to gather information regarding the driver notifications, within DRIVE C2X, the basic test scenario relies on a Roadside ITS Station, sending predefined weather situations. Passing vehicles receive these messages and, thus, present a notification while the vehicle is driving towards the situation.[69]

Hence, within the context of this thesis, multiple exemplary test rounds from two different DRIVE C2X test sides are consulted and the distance and time to the situation is evaluated by the author of this thesis in the following.

The first selected test side is the small scale test side located near O Porriño, Spain[70], where the application is stimulated within a distinct scenario. Thereby, test vehicles are driving around a small test course and passing one weather situation uniformly during every round. Since the testing route is such small, the maximum notification distance D_{max} is set to 300 m.

In Figure 7.12 the test route, the exploited weather situation, and the positions of the initial driver notification as well as the position, where the notification is removed

---

[69] A detailed evaluation of the *Weather Warning* application within DRIVE C2X is provided by [DRIVE13b] and [DRIVE14]

[70] A detailed description of all DRIVE C2X test sides is provided by [DRIVE13a].

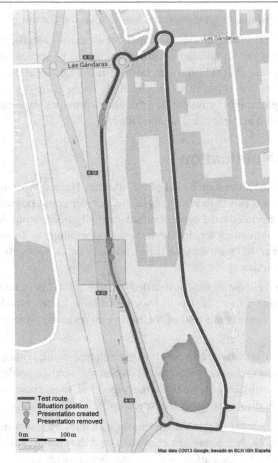

FIGURE 7.12.: Testing course with event area as well as positions of first HMI
presentation and remove of presentation within multiple test runs
near O Porriño, Spain.

from the screen, are depicted. Thereby, the positions of the first presentation and the
screen clearance, are located at almost the same position during multiple selected
test rounds. The remaining scattering is issued by the applications update interval
of about 2 s.

In accordance to the clearly clustered positions, the distances to the event d_{evt},
where the first presentation is displayed, does vary only within about 50 m during
the whole test period. The collected distances are depicted in Figure 7.13. Thereby,

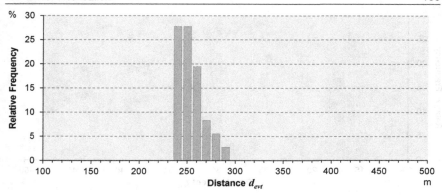

FIGURE 7.13.: Distribution of the distance to the weather event d_{evt} at first generation of a driver notification within the DRIVE C2X tests near O Porriño, Spain.

the mean distance to the event $\overline{d}_{evt}$ is about

$$\overline{d}_{evt} = 254 \, \text{m} \quad . \tag{7.19}$$

The evaluation of the time to the event t_{evt} reveals similar results. While the time t_{evt} varies between 15 s and 24 s, as depicted in Figure 7.14, the mean time to the event $\overline{t}_{evt}$ is about

$$\overline{t}_{evt} = 18 \, \text{s} \quad , \tag{7.20}$$

which is enough to react accordingly.

On the second selected test side, the large scale test side located near Tampere, Finland, the scenario is not as distinct as in Spain. In contrast, test vehicles are driving around a large test course consisting of multiple kilometers of urban roads, rural roads, and highways. During every round, vehicles are passing four weather situations from multiple angles and at different speeds. Since the testing route is large enough, the maximum notification distance D_{max} is set to 500 m.

In Figure 7.15 the test route, the exploited weather situations, and the positions of the initial driver notification as well as the position, where the notification is removed from the screen, are depicted.

Since the test course consists of a various set of scenarios, the results are not as uniform. Hence, the positions of the first presentation and the screen clearance are scattered.

In accordance, the distances to the event d_{evt}, where the first presentation is displayed, does vary due to the multiple scenarios, which are combined into one test

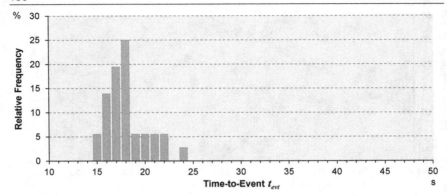

FIGURE 7.14.: Distribution of the time to the weather event t_{evt} at first generation of a driver notification within the DRIVE C2X tests near O Porriño, Spain.

round. Thereby, the distance clusters between 400 m and 500 m, but also goes down to about 170 m, as depicted in Figure 7.16. Accordingly, during the test period at the testing course in Finland, a mean distance to the event $\overline{d}_{evt}$ of about

$$\overline{d}_{evt} = 397\,\text{m} \qquad (7.21)$$

is provided during generation of the first driver notification.

The time to the event t_{evt} varies just the same. As depicted in Figure 7.17, t_{evt} is almost uniformly spread between 14 s and 35 s. However, since the minimum t_{evt} is about 14 s and the mean time to event $\overline{t}_{evt}$ is about

$$\overline{t}_{evt} = 25\,\text{s} \qquad (7.22)$$

timely driver notifications are facilitated, anyway.

Consequently, as the exemplary evaluation of the two different test courses shows, a driver notification within 15 s is facilitated in almost all cases. Hence, the proposed warning strategy relying on the recognition field does provide enough time for a driver to calmly react on a notification and adapt the driving behavior in time.

7.5. Driver Acceptance

In Car-to-X communication related research and field operational trials, a special emphasis is put on the investigation of the driver's reaction and, thus, the driver's acceptance of the C2X technology. However, each individual human reacts differently in a given scene. Thereby, the driver's, e. g., age or sex as well as personal

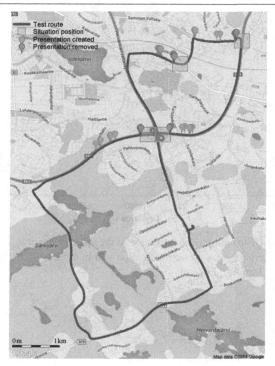

FIGURE 7.15.: Testing course with event area as well as positions of HMI presentation creation and remove of presentation within multiple test runs near Tampere, Finland.

experiences or skills have an effect on its reaction. Moreover, the behavior changes over time. Accordingly, to achieve a substantiated data basis to evaluate the driver's reaction regarding C2X presentations a huge amount of test series is required to cover the various possible drivers and scenes to some extent.

Regarding the Weather Hazard Warning application, events in field trials occur randomly, only, and may be rare. Hence, a substantiated data basis might only be achieved by consulting *driving simulations*. Thereby, each driving scene can be identically repeated frequently, with well-adjusted weather intensity and identical controlled environmental conditions, e. g., road traffic, for each individual test driver.

Accordingly, within the field operational trial of the national research project sim^{TD}, driving simulations are intensively consulted [SIMTD10c] in order to evaluate the

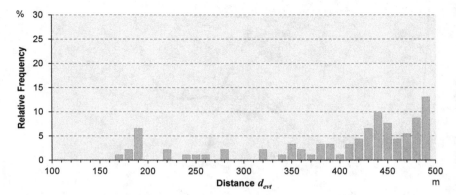

FIGURE 7.16.: Distribution of the distance to the weather event d_{evt} at first generation of a driver notification within the DRIVE C2X tests near Tampere, Finland.

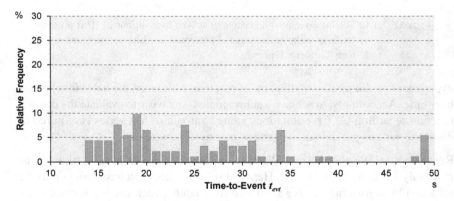

FIGURE 7.17.: Distribution of the time to the weather event t_{evt} at first generation of a driver notification within the DRIVE C2X tests near Tampere, Finland.

driver acceptance on C2X applications including the *Road Weather Warning* as detailed in Section 6.3. Therefore, the advanced driving simulator [SIMTD13d] [KGKM05], is exploited, to obtain realistic driving behavior.

Since driving speeds in urban areas are comparable low, in general, no driver reaction might be required in such scenarios at all. Thus, in driving simulation the evaluation of Weather Hazard Warning focuses on scenarios located on rural roads and highways [SIMTD10c].

Based on the instructions of the author of this thesis, suitable scenarios were developed and according tests were executed by the *Würzburger Institut für Verkehrswissenschaften* [WIVW] located at Würzburg, Germany. The results within the driving simulation in sim^TD are summarized in [SIMTD13b].

As detailed in [SIMTD13a], regarding the Weather Hazard Warning application, scenarios located on rural roads and highways are defined. While rural road scenarios cover dense fog, heavy rain, strong winds, and aquaplaning weather events, due to technical reasons, highway scenarios cover dense fog, heavy rain, and strong winds, only. In Figure 7.18 an impression of how the video screen looks like during simulations is provided.

Thereby, it has to be considered that the driving simulator does not execute the application as detailed in Section 6.3 itself. In contrast, the simulator relies on scripted scenarios that behave in the same way as the application does.

Hence, during the design of a scenario, events, e. g., a fog event, are placed at selected locations on the track. Additionally, special points with determined actions are defined, based on the schemes detailed in Section 5.2.3. Thus, if within the simulation the vehicle reaches such a specific *point on the track*, the predefined action, such as an initial driver notification, is triggered. Accordingly, presentation updates are triggered on subsequent predefined points until the position of the event is reached.

In order to estimate the effect and the acceptance of the C2X system, the reaction of a *test group* is compared with a *control group*. Thereby, the test group gets notified by the C2X system as described. In contrast, within the control group the same scenarios are executed but with no driver notifications at all. Subsequently, the results from both groups are compared against each other and against the goal of the Weather Hazard Warning application, i. e., an adapted and more safely driving behavior. Thereby, for every scenario each of both groups is set up with about 25 persons to whose up to two situations of the same event are presented.

In the following, the results of the driving simulation, regarding rural road and highways scenarios as well as the driver's impression of the C2X system are

(a) dense fog event   [SIMTD13a]

(b) heavy rain event   [SIMTD13a]

FIGURE 7.18.: Visualization of a dense fog and a heavy rain event within driving
simulation test trials.

summarized. A complete and detailed evaluation is provided by [SIMTD13a].

Rural Road Scenarios

One important factor on driving safety is the driving speed. Hence, the mean driving
speed within weather events on rural roads $\bar{v}_{rur}$ is evaluated.

Accordingly, within the test group a significant reduction by 5–10 km/h can be
observed, for all investigated weather types. In Figure 7.19 the mean driving speed
for different weather events on rural roads is displayed.

In addition, since hard and rapid braking is a threat to driving safety, the maximum
deceleration within events is evaluated. Accordingly, with exception to strong wind
events, where no significant effect can be observed, the mean maximal deceleration
on rural roads $\bar{a}_{rur}$ is reduced by 0.5–1 m/s^2 for drivers, who gets notifications in

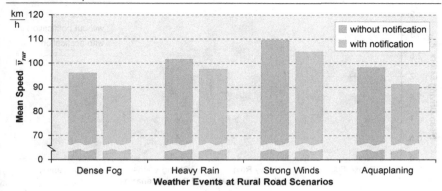

FIGURE 7.19.: The mean driving speed on rural roads $\bar{v}_{rur}$ within different weather types. For notified drivers, a significant reduced driving speed can be observed.

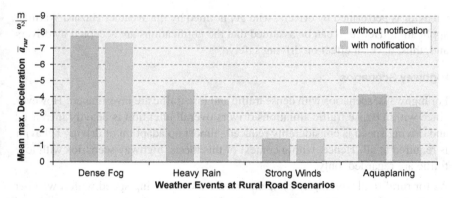

FIGURE 7.20.: The mean maximal deceleration on rural roads $\bar{a}_{rur}$ within different weather types. For notified drivers, a reduction of hard braking can be observed.

contrast to uninformed drivers. The mean maximal deceleration $\bar{a}_{rur}$ for all weather events is depicted in Figure 7.20.

Finally, the drivers are evaluated by observers regarding the driving safety. In Figure 7.21 the mean number of driving errors per driver on rural roads $\bar{e}_{rur}$ is depicted for multiple event types. Accordingly, a by 0.5–1 reduced mean number of driving errors occur for notified drivers, which might heavily increase road safety.

Summarizing, regarding driving speed, deceleration, and number of driving errors, in contrast to the control group, a considerably adapted and smoother driving

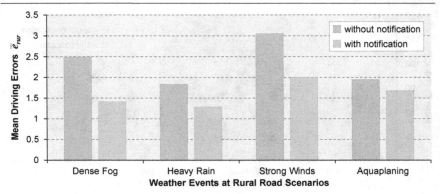

FIGURE 7.21.: The mean number of driving errors on rural roads $\bar{e}_{rur}$ within different weather types. For notified drivers significant less errors can be observed.

behavior is performed by drivers, who get notified by the Weather Hazard Warning application. Hence, it can be assumed that the notifications are accepted by drivers and, thus, have an effect on driving safety.

Highway Scenarios

For highways, scenarios with dense traffic and lose traffic are investigated. However, since within dense traffic situations drivers overall attention is heavily increased and driving speed is considerably reduced, almost no adaption of driving behavior is required at all. Hence, within context of this thesis, highway scenarios with lose traffic are regarded, only.

As for rural roads, on highway scenarios the mean driving speed within weather events $\bar{v}_{hwy}$ is evaluated, too. Accordingly, the mean driving speed is reduced by 5–10 km/h within weather situations of all investigated types. As depicted in Figure 7.22, especially for dense fog events a significant reduction can be observed.

To evaluate the driver's braking behavior, the mean percentage of critical braking on highways $\bar{b}_{hwy}$, i. e., decelerations above -4.5 m/s^2, is investigated. As depicted in Figure 7.23, the frequency of critical braking is reduced by about 0.5–1 percentage points within the test group. Hence, drivers within the control group compromise the driving safety by brake significantly more spontaneously and rapidly.

Moreover, especially on highways, the distance to the vehicle ahead is of substantial importance for driving safety. However, the evaluation on the mean minimal distance to the vehicle ahead on highways $\bar{d}_{hwy}$ shows that the mean minimum

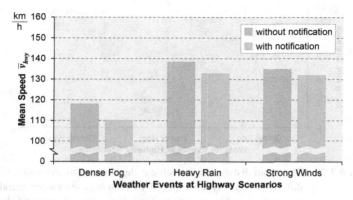

FIGURE 7.22.: The mean driving speed on highways $\bar{v}_{rur}$ within different weather types. For notified drivers significant reduced driving speed can be observed.

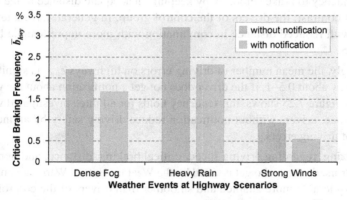

FIGURE 7.23.: The mean percentage of critical braking on highways $\bar{d}_{rur}$ within different weather types. A critical braking is a deceleration above $-4.5\,\text{m/s}^2$. Especially within fog and rain events a considerable smoothed braking behavior can be noticed.

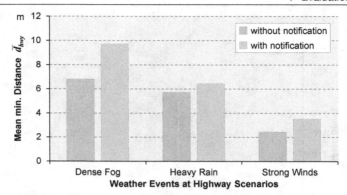

FIGURE 7.24.: The mean distance to the vehicle ahead on highways $\overline{d}_{rur}$ within different weather types. Especially within fog events a considerable increased distance can be observed, if the driver is notified about the event.

distance within weather events is significantly increased, if the driver is notified about dangerous events. As depicted in Figure 7.24, especially for dense fog, $\overline{d}_{hwy}$ is increased by almost 2 m,

Consequently, regarding the evaluation of braking and distance, notified drivers have a tendency to brake smoothly by keeping an adequate distance to the vehicle ahead. In contrast, the drivers of the control group, in general, have a tendency to drive with reduced distance in combination with more unpredictable braking maneuvers.

Accordingly, the mean number of driving errors on highways $\overline{e}_{hwy}$ is significantly increased by about 0.5–1, if the driver does not get a notification about the weather event. As Figure 7.25 shows, this tendency holds for all three investigated weather types. Hence, drivers without notification reduce driving safety by an increased number of driving errors.

Summarizing, regarding driving speed, critical braking, distance, and number of driving errors, drivers, who get notified by the Weather Hazard Warning application drive considerable more adapted and smoother than drivers of the control group. Hence, notifications are accepted by drivers and, thus, have an effect on driving safety on highways.

Driver Impression

As the evaluation on scenarios located on rural roads and highways indicate, drivers have a tendency to take notifications provided by a C2X system seriously. Accord-

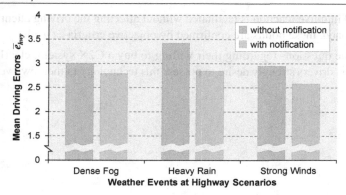

FIGURE 7.25.: The mean number of driving errors on Highways $\bar{e}_{hwy}$ within different weather types. For notified drivers significant less errors can be observed.

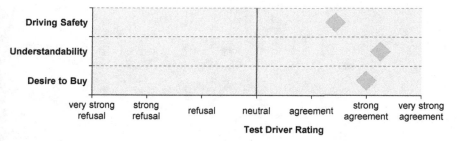

FIGURE 7.26.: Impression of the C2X system as reported by the test drivers of the driving simulation. The respective questionnaires consider personal impression of driving safety, the understandability of the provided notifications, and the personal desire to buy and possess such a C2X system.

ingly, they drive slower, smother, and more attentive due to objective criteria.

However, their personal impression regarding the C2X system is investigated within driving simulation, in addition. Hence, questionnaires include a variety of questions regarding their personal impression of driving safety, the understandability of the provided notifications, as well as their desire to buy and possess such a system are captured.

As depicted in Figure 7.26, the drivers agree up to strongly agree that the notifications increase the driving safety. In addition, due to the impression of the test drivers, the notifications are very clear and easy to understand, such that an appropriate and intuitive reaction is facilitated without affecting the driving attentiveness. As mentioned, these points are confirmed by objective criteria as well.

Finally, the questions regarding their willing to buy a C2X system, clearly show that the test drivers strongly desire to possess this technology in their own vehicles.

8. Summary and Conclusion

An overall concept of a road weather related application for Car-to-X communication was presented within this thesis. Thereby, the underlying C2X system architecture, vehicle and infrastructure stations, utilized communication frequencies, and C2X messages are introduced and communication concepts, such as message forwarding and pseudonym changes, were detailed.

By using C2X communication technology, the developed *Weather Hazard Warning* application overcomes three major information gaps by

1. regarding information about local weather conditions provided by infrastructure-based services relying on, e. g., a large network of advanced weather measurement stations,

2. sharing knowledge about weather events, detected by vehicle's local sensors with other vehicles in the local area, and

3. gathering weather related data on a vehicle's way along the road and providing these sensor data to infrastructure-based services to enlarge their knowledge about current weather conditions.

In order to facilitate the application, by regarding and enhancing previous works, schemes for detecting current weather conditions within vehicles were developed. With these schemes, weather situations such as dense fog, heavy rain, aquaplaning, heavy snowfall, and slippery roads due to ice or snow might be recognized by relying on sensors available in current vehicles, only. Additionally, an outlook on infrastructure-based detection and on how situation detection and detection reliability might be further enhanced by regarding additional sensors, was provided.

Information within C2X communication in general and especially weather related data are closely associated with their location. In order to efficiently handle location information, a novel concept to transform geographical coordinates into a local Cartesian coordinate system was developed. Especially within automotive control units, where computation resources are hard limited, such a transformation relying on a few operations only provides huge advantages. Additionally, novel and highly efficient approaches facilitating operations on rectangular areas, as consulted within C2X communication, were presented. This especially includes algorithms to determine a rectangle surrounding a trace, to calculate the distance from an arbitrary point to the nearest border of a rectangle, to verify if a point is located inside a rectangle, and to calculate the overlapping area of two rectangles. These approaches, regarding coordinate transformation and handling of rectangular areas, are neither limited to the context of the Weather Hazard Warning application nor to C2X communication.

In order to notify the driver about dangerous situations, strategies to merge and maintain weather related information from multiple sources were detailed. Hence, an approach to determine which individual reported situations belong to the same overall weather event was detailed. Thereupon, a concept for driver notification by means of consulted media as well as notification time and intensity was developed and discussed. Therefore, notification levels and the recognition field were introduced.

Since C2X communication is vulnerable to various attacks on information's integrity and driver's privacy, multiple countermeasures were developed, in addition. Security considerations and further countermeasures are detailed in Appendix A. Developed countermeasures especially include a novel approach for mobility data verification, which consults a *Kalman filter*-based tracker to predict and verify movements of adjacent vehicles. Since mobility information is used by almost all C2X components and applications, the verification might be applied on lower communication layers. In order to additionally provide countermeasures even for the application layer, verification strategies for weather related data by means of validity time aware reliability adaption and concepts for cooperative information confirmation were presented.

Within the German research project sim^{TD}, the author of this thesis co-developed the respective Car-to-X large scale field operational trial system architecture and security solution. Additionally, in order to verify and evaluate developed approaches, the concept of the Mobility Data Verification and adaptions of the Weather Hazard Warning application were fully implemented for and deployed within the field operational trials of the national and European research projects sim^{TD} and DRIVE C2X, respectively. Hence, the carried out evaluation did not rely on simulations only, but was based on real word scenarios and a huge amount of real world data, instead. Thereby, efficient calculations and data management as well as a reliable situation detection and driver notification were observed.

The evaluation of driver acceptance additionally showed that people indeed adapt their driving behavior according to the presented notifications. Accordingly, the goal of a significant reduction of critical driving situations and driving errors was verified. Moreover, test drivers consistently expressed their anticipation to use a C2X system in general and the Weather Hazard Warning application especially in their own prospective vehicles.

Consequently, towards next steps of market launch, the use-case of the Weather Hazard Warning is selected as a *Day-One* application, i. e., the set of applications that are planned to be launched within the first available systems. The respective specification of triggering conditions within the C2C-CC is partially taken over from the developed scheme for weather situation detection. Hence, it serves as basis for the European standardization of Car-to-X applications.

For future development, C2X communication will facilitate far more than driver notification, as implemented within context of this thesis, and might be integrated into common platforms, e. g., the *Automotive Open System Architecture* (AUTOSAR) [AOSA] [KBFL+11], instead of Java/OSGi. Next steps may provide driving interventions, like pre-braking, and driving assistance. Moreover, within far future, inter vehicle communication will be a prerequisite for platooning or autonomous driving.

However, for driving interventions and automated driving a lot of research is still required. Especially, information quality and reliability have to be improved, such that higher *ASIL* levels might be ensured. Towards these visions, developed data verification approaches and techniques to handle geographical accurate rectangular areas are sustainable advancements.

Regarding upcoming market launch of C2X communication systems, an *introduction dilemma* has to be conquered. The majority of C2X applications rely on a certain penetration rate to operate reliable and trigger dependable driver notifications [C2C07]. Assuming that every new middle class and above vehicle, i. e., about every second vehicle, is equipped with C2X technology, it will take about 3 years to reach 10 % and more than 15 years to reach nearly 50 % penetration rate [SMM+05]. Even in the optimal case, if, from now on, every new vehicle will be equipped, it will take up to 1.5 years to reach 10 % and more than 7 years to reach 50 % penetration rate. Consequently, further solutions, like dedicated nomadic devices to be attached into existing vehicles [RITA10], or hybrid approaches involving mobile phones [ZSGW09] or even satellite communication [KSSH12] are under discussion.

Moreover, for cooperative applications, e. g., the use-case of the Intersection Collision Warning, both involved vehicles have to use C2X communication. Hence, at a penetration rate of 50 %, there is just a 25 % chance for the use-case to work properly.

Thus, even in optimistic cases, car manufactures will be faced with the huge challenge to sell expensive C2X equipment that will have almost no functionality for more than 10 years, i. e., nearly the expected lifetime of the vehicle. Consequently, to facilitate an appropriate penetration rate in foreseeable future, introduction of C2X systems has to be enforced by vehicle safety institutes, e. g., Euro NCAP, and regulated by law.

However, despite all known obstacles, a first car manufacturer has already announced a vehicle featuring C2X communication within 2017 [SCN14]. Even if this vehicle will not be sold at high volumes, it is a first step to establish C2X communication on the roads.

A. Security Considerations

Car-to-X communication is considered to be one of the most promising attempts to improve active safety and traffic efficiency in the near future. However, such advanced vehicle communication systems are vulnerable to several attacks against security and driver's privacy.

Especially safety critical applications such as, e. g., *Forward Collision Warning* or *Intersection Collision Warning*, require instant driver reaction. Hence, corrupted or forged messages may cause wrong driver actions and therefore have fatal consequences for road safety. However, even to safety applications that are not time critical, like the *Weather Hazard Warning*, falsified injected messages at least lead to low acceptance of C2X technology by the customer. Additionally to injected forged messages, especially following vehicle movements, i. e., tracing the identifiable driver by means of eavesdropped messages may be an issue for driver's privacy protection.

In the following possible attacker models, specific attacks, and private issues are discussed and multiple appropriate countermeasures investigate or actually designed by the author of this thesis are detailed.

However, successful attacks wills never be prevented entirely. Instead, each countermeasure may increase the effort required to apply an attack, such that the number and probability of feasibility attacks is decreased.

A.1. Attacker and Threat Analysis

Despite all aspired benefits, applications based on Car-to-X communication will contribute, such systems are highly vulnerable towards attacks. This section outlines possible attacker types and lists exemplary threats to C2X applications. Comprehensive analysis of attacks against C2X communication is carried out, e. g., in [BH07] or [RH05][71]. In context of this thesis three major classes of possible threats to C2X

---

[71] Additionally, extensive analysis not limited to C2X communication, regarding also psychological profiles and attacker backgrounds are, e. g., detailed in [CCD08] or [ECK13].

communication are given by either *Denial of Service* attacks, *Message Injection*, or *Privacy Infringement*.

Denial-of-Service (DoS) attacks thereby focus on disabling C2X communication at all. Therefore two major approaches can be identified. Firstly, the communication channel is flooded with data such that valid messages might not be successfully transmitted. These data not necessarily have to be correct C2X messages, instead generating *white noise* at high amplitude will effectively disable any communication. On the other hand, receiving ITS Stations may be utilized up to their full computation capacity by spamming random, but syntactical correct, messages. Hence, the C2X system is overstressed while interpreting and processing the data. These attacks require only sparse technological equipment and low understanding of the C2X communication technology, but are capable to achieve huge effects.

In contrast, *Message Injection* refers to more sophisticated and directed attacks. Thereby, valid C2X messages are injected to either causing disturbance by driver misleading or exploiting the system for own advantage. Message injection includes replaying of messages, manipulating of existing messages, or generating of completely forged message. To facilitate suchlike attacks, advanced knowledge about functionality of C2X applications and comprehensive equipment, especially availability of valid C2X certificates, is required. Accordingly, detection of such an enhanced adversary is a challenging task requiring sophisticated countermeasures.

Finally, due to safety purposes, vehicles constantly broadcast their current position and additional data as well as unique identifiers within messages. These messages may be observed by an adversary, exploiting them to retrieve driver's movements[72]. Regarding that almost each vehicle may be mapped to one person, as detailed in Section 2.4, identifying a vehicle is approximately identical to identifying an individual person. This represents a huge *Privacy Infringement*. However, in general it is recommends to not include more information into broadcasted C2X messages, than really needed by the receiving application [SMK09].

Observing C2X communication generally occurs without any possibility to be noticed. This complicates applying of specific countermeasures. Such surveillance typically requires a comparable huge effort, since messages have to be gathered on multiple locations. However, even the knowledge that a person visits just a few specific locations may reveal private information. Moreover, privacy issues have to be regarded within C2X communication, such that legally demands are fulfilled and customer acceptance of C2X technology is increased, additionally.

---

[72] Even with just few and vague available data, this poses a significant threat to the user's privacy and discloses confidential information as insistently detailed in different but comparable context in [FAZ10].

A.1.1. Hardware Data Intrusion

To complicate *Message Injection* by exploiting original C2X technologies, also vehicular on-board network has to be protected from modification to ensure trustworthiness and security. To face these challenges, within, e. g., the European research project EVITA [EVITA] possible attacks on vehicular on-board networks are complicate [HAR+09] and an architecture for secure automotive on-board networks is designed.

Nevertheless, protecting on-board systems of every ITS station against manipulation, message injection, or modification is a very challenging and complex task. Moreover, to overcome the *introduction dilemma* at early introduction phase of C2X communication systems hybrid approaches which engage mobile phones or dedicated nomadic devices will be present [RITA10]. Both may introduce certain additional security weaknesses.

On the one hand security credentials stored in such devices may be more easily extracted than from sophisticated embedded vehicular on-board security hardware. Even with the use of trusted platform modules extraction of confidential information, e. g., secret keys, is still possible [TAR10]. On the other hand, interfaces to vehicle on-board networks, e. g., CAN bus, as needed for diagnostics, nomadic devices, or connecting mobile phones may be easily exploited to insert faked information into vehicle modules [KCR+10].

In fact, it has to be assumed that an attacker is capable to generate valid C2X messages generated by original C2X communication systems. This attacker sends valid signed messages, since he may inject data directly into the vehicles CAN or is in charge of a valid certificate. Such an attacker is hardly detectable.

Consequently, especially during introduction phases of C2X systems, the injection of bogus messages cannot be fully prevented. Thus, countermeasures based on cryptographic approaches, only, as detailed in Section A.3, are not sufficient for a reliable protection of C2X communication. Accordingly, in addition to protecting communication links between ITS stations by means of signatures and encryption, verification of message content, as detailed in Section 4.1 and Section 4.2, is necessary and, hence, presented by the author of the thesis.

A.1.2. Vehicle Tracking

One of the basic mechanism facilitating safety related applications in C2X communication, is the periodically broadcast of vehicle position, speed, heading, and additional vehicle specific information. As detailed in Section 2.3.1 these mobility data are sent at least every second and up to ten times a second. Due to the cooperative and broadcast nature of C2X communication, such basic messages are readable for every ITS station within communication range.

Since these messages contain multiple unique identifiers, e. g., the *Station ID* or *MAC Address*, and, moreover, are digitally signed within a *Public Key Infrastructure* (PKI), all messages can unambiguously be mapped to an originating vehicle. Thus, any vehicle path can be observed and followed. Furthermore, by collecting even partial samples of these messages and omitting identifiers, the entire path can easily be retrieved [GH05].

Inferring of such sensitive user data is a serious issue for driver's privacy. Hence, tracking of vehicles has to be impeded by means of legal restrictions and technical countermeasures. In fact, comprehensive integration of privacy techniques is a prerequisite for a successful deployment of C2X systems.

Potential *Privacy Infringements* based on vehicle tracking are outlined within the description of attacker models in the following. Additionally, possible countermeasures to ensure an acceptable level of privacy protection are discussed in Section 2.4.

A.1.3. Roadside Attacker

Due to its low complexity, a stationary attacker located within communication range on the roadside is identified as the most likely adversary [LSS+08]. Just by exploiting adequate software on a simple laptop equipped with an antenna tuned to 5.9 GHz harmful interactions are facilitated.

Regarding privacy issues, while gathering and observing communication emitted from vehicles, visited places and movements of drivers are revealed. This already has a huge impact on a person's privacy. Moreover, by applying comprehensive traffic surveillance, automated ticketing due to traffic or parking violations is enabled or adapted rates to insurances may be facilitated. Additionally, knowing the precise location of potential customers, targeted marketing is facilitated by adjusting advertisements regarding the driver's background. Applying more criminal intentions, e. g., daily routines, such as a person's typical times for leaving home for work, may be profiled automatically to estimate possible time-slots for burglars.

Countermeasures to this may be provided by changing identifiers as detailed in Section 2.4 or by restricting the radiation of C2X messages as detailed in Section A.2.

In contrast to these rather targeted but passive adversaries, active interaction originated from the roadside may be most likely occur by "*script kiddies*" [LSS+08].

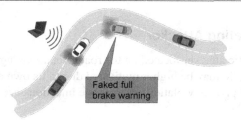

FIGURE A.1.: Probably attack on *Forward Collision Warning* application, with a faked full brake at a low visible position.

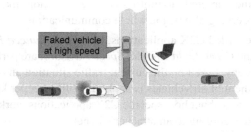

cf. [JBSH12]

FIGURE A.2.: Probably attack on *Intersection Collision Warning* application, with a faked priority vehicle approaching the intersection at high speeds.

These attackers unlikely have a deeper understanding of the technology and might be motivated just by own ambitions, curiosity, and their play instinct instead of the goal to seriously harm other persons [CCD08]. However, even by incorrect use of exploited equipment Denial-of-Service attacks are facilitated.

Additionally, more sophisticated and targeted attacks are imaginable. By interfering the *Forward Collision Warning* with a faked full brake warning at a low visible position as illustrated in Figure A.1 may cause affected drivers to instantly perform a powerful brake. Accordingly interacting with the *Intersection Collision Warning* by indicating a faked priority vehicle approaching the intersection at high speeds, as Figure A.2, leads to similar results. Both attacks might increase the probability of rear-impact collisions significantly and, hence, may cause substantial harm.

Countermeasures to suchlike active disturbance are, e. g., given by restricting the antennas reception range as detailed in Section A.2, by approaches complicating generation of valid message by exploiting digital signatures as stated in Section A.3, or by extensive verification of message content as detailed in Section 4.1 and Section 4.2.

A.1.4. Participating Attacker

In contrast to a static attacker located on the roadside, a moving adversary participating the road traffic may be higher motivated due to its own advantage. Hence, rather professional privacy violations or message injections become very likely to occur.

By evaluating C2X message from a long distance, e. g., a private investigator might observe and follow person way easier without the risk of getting noticed or losing the victim. In fact, the needed effort required for surveillance is decreased. Especially, such rather targeted monitoring is easier and more likely to occur than attacks based on observing almost the whole communication.

Traffic efficiency regarded C2X applications, like the *Enhanced Route Guidance and Navigation* or the *Green Light Optimal Speed Advisory*, are performing adapted traffic control. Thereby, based on the current traffic situation an optimized traffic flow with fair treatment of all vehicles is applied. With a deeper knowledge about the C2X technology and about how specific C2X applications work, such use-cases might be influenced according to an attacker's benefit.

Hence, by generating, e. g., faked *Traffic Jam Warnings* regarding own aspirated route, other vehicles are redirected by their adapted navigation. Thus, a free and uncongested road is achieved for the attacker.

Accordingly, by actively injecting, e. g., multiple vehicles placed on the own line, an adversary may manipulate the system to obtain green traffic lights as depicted in Figure A.3. To complicate such exploitation of C2X communication use-cases, in-depth analysis of message content as detailed in Section 4.1 or exemplary shown in Section 4.2 may be applied.

Analogous, an adversary might pretend to be a public transportation vehicle or even an emergency vehicle in order to fool a *Public Transport Prioritization* or an *Emergency Vehicle Prioritization* application. This way, a prioritized traffic flow for the own driving lane may be achieved. However, countermeasures to these attacks are effectively given by digital signed messages with according certificates as detailed in Section A.3.

A.2. Secure C2X Beamforming

In Car-to-X communication appropriate security mechanisms have to be applied on every communication layer, even on the very first, i. e., already during sending and receiving of messages.

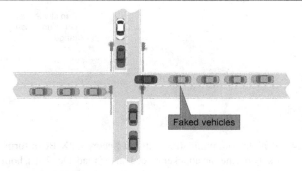

cf. [JBSH12]

FIGURE A.3.: Probably attack on *Green Light Optimal Speed Advisory* application,
with a multiple faked vehicles causing traffic lights to give priority
to the attacker.

Respective security related standardization, i. e., IEEE 1609.2 [IEEE06], mainly
focuses on pseudonym changes as well as on cryptographic methods based on a
Public Key Infrastructure, i. e., message signing and encryption. These techniques,
take effect on the *Networking & Transport* layer and upwards, leaving medium
access unconsidered, by design. To overcome this lack of security, *Secure C2X
Beamforming*, is introduced in [SSH09] to provide security techniques on the
physical layer.

Within Secure C2X Beamforming the physical propagation and reception of transmitted radio signals is modified and adapted based on an antenna-array. By means
of radiation pattern control antennas focus transmitted power towards a desired
direction. In addition, by utilizing beamforming ITS stations may improve signal
reception towards intended and rejecting unwanted directions. Thereby, not only
the exchanged data are secured, but also the medium, on which an attacker can
operate, is restricted. Hence, both goals, secure communication with respect to the
driver's privacy needs, are reached at the same time.

As previously detailed, an attacker placed on the roadside, injecting faked messages
or eavesdropping communication, is most likely. Conventional wireless communication basically covers a radial area around the antenna. Figure A.4 depicts such
a situation, where the attacker on the roadside is within the radial communication
range of the vehicle driving along the road. In contrast, Figure A.5 illustrates a
vehicle exploiting Secure C2X Beamforming to restrict the communication range
mainly to the road. Hence, an attacker placed on the roadside may not be able to
interact with the vehicle and, hence, not to inject or to eavesdrop C2X messages.

In [SSH10b] and [SSH10a] a C2X Beamforming antenna array is presented. This

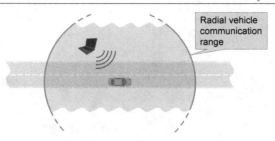

cf. [SJ10]

FIGURE A.4.: Vehicle communication without Secure C2X Beamforming most
likely reaches an attacker placed on the roadside. Thus, both falsified
messages can be injected and communication can be eavesdropped.

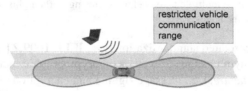

cf. [SJ10]

FIGURE A.5.: Vehicle communication with Secure C2X Beamforming unlikely
reaches an attacker placed on the roadside. Thus, neither falsified
messages may be injected nor communication may be eavesdropped.

antenna, developed in [MIG09] and [RUP11], is suitable for most of the safety related use-cases as defined by the *Car-to-Car Communication Consortium* (C2C-CC)
[C2C07]. It consists of two orthogonal antenna arrays which are steered independently from each other [SJ10] such that a great variety of different transmission
patterns is given [BAL05].

A.2.1. Radiation Patterns for the Weather Hazard Warning

Within this section, the author of this thesis suggests specific radiation patterns
suitable to be consulted within the Weather Hazard Warning application.

As detailed in Section 1.2, within *Weather Hazard Warning* application, different message types in different communication scenarios are transmitted between
vehicles and between vehicles and infrastructure. Accordingly, three basic communication scenarios are identified by the author of the thesis.

Firstly, the *Weather Hazard Notification* scenario. Thereby, DENMs are sent
initially from vehicles to other vehicles or to central stations via RISs. Such a
communication scenario is illustrated in Figure A.6.

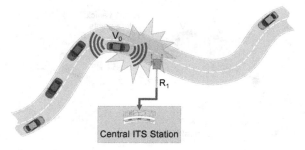

cf. [JH11]

FIGURE A.6.: The *Weather Hazard Notification* communication scenario, where a
detecting vehicle V_0 notifies surrounding vehicles and, via a roadside station R_1, a central stations about the hazard.

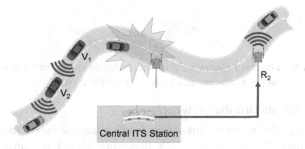

cf. [JH11]

FIGURE A.7.: The *Weather Hazard Forwarding* communication scenario, where
an initial sent message is forwarded by receiving vehicles V_1 and
V_2 as well as by a central station via a roadside station R_2.

The second scenario refers to the *Weather Hazard Forwarding*, whereas *Store &
Forward* or *Keep-Alive Forwarding* mechanism [C2C07] [ETSI13c] are advocated,
respectively. Hence, already sent messages are distributed via multiple hops over
larger distances or retained over longer times within the distribution area. In
Figure A.7 forwarding over several hops, different lanes, and via RIS is depicted.
Forwarding, vehicles may send messages into opposite direction of receiving, i. e.,
usually away from the center of the respective hazard, only.

The last scenario, the *Probe Data Distribution*, applies to the transmission of
PVDMs from vehicles to central stations via a RIS. As mentioned afore, the
exchange of PVDMs between vehicles is not intended. However, it has to be
considered that PVDMs are large messages, which causes comparatively long
transmission times. Hence, the relative position between vehicle and RIS will
vary considerable during transmission. In Figure A.8 a communication scenario

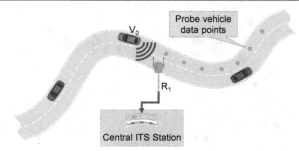

cf. [JH11]

FIGURE A.8.: The *Probe Data Distribution* communication scenario, where the vehicle V_0 transmits collected prove vehicle data via a roadside station R_1 to a central station.

cf. [SJ10]

FIGURE A.9.: The *Road-Restricted Pattern*, restricting the radiation to front and back whereas the area beside the road is avoided.

regarding Probe Data Distribution is depicted.

In the following, the author of this thesis details adequate radiation patterns out of the set facilitated by the Secure C2X Beamforming antenna-array [SSH10b] that fulfill the requirements of the three identified communication scenarios of the Weather Hazard Warning.

The Weather Hazard Notification scenario applies whenever the application detects a hazard such as black ice, heavy rain, or dense fog. Hence, a corresponding message is instantly distributed to all surrounding vehicles. Since weather hazards are not limited to the vehicles positions only, but rather have an effect on the entire area around, these DENMs are relevant for all vehicles in communication range.

Consequently, an a priori restriction of the transmission area is not anticipated for this first scenario. Hence, the *Road-Restricted Pattern* as illustrated in Figure A.9 may be applied. Nevertheless, a radiation adaption according to the actual position of vehicles in communication range is intended. Accordingly, also an *Endfire Pattern* or a *Frontfire Pattern* as depicted in Figure A.10 may be suitable for this communication scenario, if receivers are located in one direction only.

In case of the Weather Hazard Forwarding scenario, messages are forwarded over large distances and over longer times. In contrast to the previously detailed communication scenario, within this scenario messages are sent before at least once.

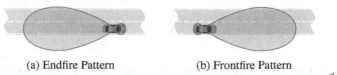

(a) Endfire Pattern   (b) Frontfire Pattern

cf. [SJ10]

FIGURE A.10.: The *Endfire Pattern* restricting the radiation to the back and the *Frontfire Pattern* restricting the radiation to the front, respectively, whereas the area beside the road is avoided in both cases.

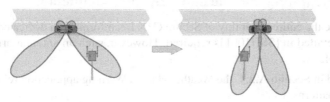

FIGURE A.11.: The *Position-Based Pattern*, focusing the radiation beam statically or dynamically towards a specified position.

Hence, the message may not be intended to be sent to all surrounding vehicles. Consequently, the radiation pattern may be adapted in a way that already notified vehicles left out.

Accordingly, especially by forwarding a message, regarding the respective intended dissemination direction of the message, an *Endfire Pattern* or a *Frontfire Pattern* is most suitable.

The *Probe Data Distribution* communication scenario refers to a data transmission from a vehicle to a roadside station. This requires the transmission of large data sets when the vehicle is passing by a RIS. Hence, with dynamic vehicle positions, a mutual beamforming of both, sender and receiver, is anticipated. Consequently, in order to support a secure transmission the *Position-Based Pattern* with dynamic beam steering as depicted in Figure A.11 is appropriate for this communication scenario.

Bundled radiation characteristic, as applied within Secure C2X Beamforming, has an effect not only on transmission but also on signal reception. Hence, areas wherefrom an attack, e. g., jamming the communication channel or injected faked messages, is originated may be excluded by means of adapting radiation beams. Thereby, either an adequate pattern, not including the attacker's position, may be selected, or the radiation power into the attacker's position may be adjusted as depicted in Figure A.12. Particularly for *Weather Hazard Notification* and *Weather Hazard Forwarding* scenarios, this approach is applicable.

cf. [SJWH11a]

FIGURE A.12.: By adjusting the radiation beam, known attackers are excluded while communication with other road participants is still enabled.

A.2.2. Integrating into the sim^{TD} System Architecture

To facilitate the entire capability of Secure C2X Beamforming, a dedicated antenna-array as detailed in [SSH10b] is required. However, such an antenna-array is not yet available.

As detailed in Section A.2.1, the Weather Hazard Warning application relies on few radiation patterns, i. e.,

1. the *Road-Restricted Pattern*, restricting the radiation to front and back,

2. the *Endfire Pattern*, restricting the radiation to the back,

3. the *Frontfire Pattern*, restricting the radiation to the front, and

4. the *Position Based Pattern*, focusing the radiation dynamically towards a specified position,

provided by the Secure C2X Beamforming antenna-array, only.

As outlined in Section 6.1.2, the sim^{TD} antenna module includes two 5.9 GHz antennas, one focused to the front and one focused to the back. With such an antenna module, not the entire variation of patterns provided by a dedicated antenna-array may be created. However, by controlling each of the two antennas individually, at least the first three patterns applicable for the Weather Hazard Warning application may be reproduced.

As mentioned in [SIMTD09k], the power level of each of the two 5.9 GHz antennas of the sim^{TD} antenna module may be dynamically adjusted from 0 dBm in steps of 0.5 dB up to 21 dBm. Thus, possible attacker positions may be excluded in addition.

Hence, in [SJWH11b] and [SJWH11a] a preliminary approach for integrating Secure C2X Beamforming into an existing C2X system architecture is demonstrated. In this section the concept is outlined, refined, and improved.

To integrate Secure C2X Beamforming into the sim^{TD} system architecture, three components are introduced.

1. The *Secure Transmit*, controlling the transmission power,

2. the *Secure Receive*, controlling the reception sensitivity of the antennas, and

3. the *Trust*, evaluating and storing information about neighbored vehicles.

The sim^TD *Security* system component evaluates messages by means of digital signatures. Based upon this, the *Trust* sub-component may evaluate individual neighbored vehicles as untrustworthy or attacker. Moreover, the *Mobility Data Verification* component as detailed in Section 4.1 is part of the sim^TD architecture as detailed in Section 6.2 and may provide additional plausibility information to be regarded by the Trust sub-component. The evaluation result and position for each neighbored vehicle is provided to both, the *Secure Transmit* and the *Secure Receive* sub-component by the Trust sub-component.

Based upon the radiation pattern applicable for the current message, as detailed in Section A.2.1, the front, the back, or both antennas may be activated for transmission. By regarding the own vehicle's position and the positions of neighbored vehicles, the transmission power for the respective antenna may be further refined to the actual distance. The required own position is provided by the Positioning component, i. e., the VAPI Server, whereas the position of neighbors may be provided by the C2X Network component, where they are stored to determine C2X message forwarding. Moreover, based on the data provided by the Trust sub-component, the transmission power may be further adjusted to exclude known attackers at sending. Hence, the required *transmission power* for each of the two antennas may be calculated and included into the message and, thus, adopted by the ITS-G5A Access during sending.

On the other hand, the Secure Receive may adjust the *reception sensitivity* directly at the ITS-G5A Access. The reception sensitivity thereby is refined according to the *Sensitivity Adaption Loop*. Hence, based on the position of a known attacker, for each antenna it is individually evaluated if the attacker is within the current communication range. If the attacker is outside the range, the sensitivity may be increased by 0.5 dB. If it is inside the communication range the sensitivity is decreased by 0.5 dB. The resulting reception sensitivity is outputted and the loop is repeated after a waiting time t_{rep} of, e. g.,

$$t_{rep} = 100\,\text{ms} \quad . \tag{A.1}$$

The *Sensitivity Adaption Loop* is depicted in Figure A.13.

The integration of the Secure C2X Beamformingin into the overall sim^TD system architecture is depicted in Figure A.14. Thereby, described introduced components, new communication links, and added information required in addition to the deployed field operational trial system are highlighted.

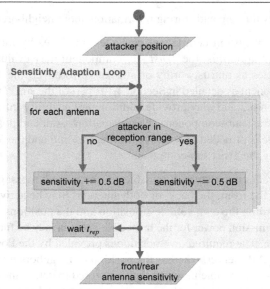

cf. [SJWH11a]

FIGURE A.13.: The *Sensitivity Adaption Loop* of the Secure C2X Beamforming, adapted for the sim^{TD} field operational trial system architecture.

A.3. Cryptographic Security

In IEEE 1609.2 [IEEE06] countermeasures by means of cryptographic techniques based on *Elliptic Curve Cryptography* (ECC) are proposes. The standard specifies *digital signatures* according to ECDSA [NIST09] to authenticate exchanged C2X messages. Additionally, to ensure confidentiality, optional *encryption* according to ECIES [IEEE04] is specified. For both, signatures and encryption, a *Public Key Infrastructure* (PKI) has to be established to generate certificates by *Certificate Authorities* (CA).

Cryptographic security is mapped to the *C2X Communication Service*, which handles incoming and outgoing messages. Hence, without the need to interpret the message contends, messages can be signed, verified, encrypted, or decrypted. Moreover, mechanisms to update vehicles certificates via a PKI can be applied directly without involving unrelated systems components.

With approaches provided by cryptographic security, several goals, e. g., according to [ECK13], are reached. Within context of this thesis, these are adapted to the C2X communication domain as follows.

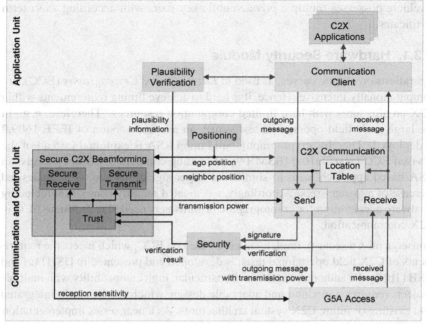

cf. [SJWH11a]

FIGURE A.14.: Integration of the Secure C2X Beamforming into the sim^TD field operational trial system architecture with respect to components and communication links.

- *Sender Authentication* is applied to every C2X message by verifying the attached signature. Thus, it is ensured that the originating ITS station owns valid public/private keys issued by a known Certificate Authority.

- *Sender Authorization* ensures, if required, that a C2X message is originated only by an ITS stations that is permitted to send the respective message, e. g., an *Emergency Approaching Warning* message. Hence, it is verified that the according privileges are set in the corresponding certificate.

- *Message Integrity* is verified by including and evaluating hashes on message's header and content. In doing so, manipulated messages are recognize.

- *Message Confidentiality* guarantees that no one but the intended receiver may read the message's content. This is not required by all C2X messages, but may be achieved by encryption.

Thereby, it has to be regarded that digital keys and related certificates have to be changed frequently for privacy reasons as further detailed in Section 2.4. Hence, a vehicle possesses multiple private/public key pairs with according short term certificates.

A.3.1. Hardware Security Module

Operations on elliptic curves, as used in *Elliptic Curve Cryptography* (ECC), are computationally intensive. Hence, it is hard to achieve timing requirements within C2X architectures with hard limited computation resources. Therefore, e. g., in the large scale field operational test sim^{TD}, an adapted version of IEEE 1609.2 is deployed. Thereby, cryptography build upon RSA is exploited instead of the advised ECC [SIMTD10a] [BSM+09]. Providing comparable security levels, an RSA-based approach leads to significant larger signatures and, thus, undesired increase of message sizes. Accordingly, a dedicated security component including hardware acceleration for cryptographic primitives is needed to facilitate ECC in C2X communication.

Consequently, a security implementation based on ECC, which meets the requirements of C2X field operational tests, is developed and presented in [JSH11a] and [JSH11b] by the author of this thesis. In particular, high compatibility with multiple projects requires a modular and adaptable design, which can be easily integrated into existing or future C2X system architectures. As a near series implementation, it further has to be compliant to the IEEE 1609.2 standard.

Consequently and according to standardization, the following five major tasks has to be performed by the developed *Hardware Security* module.

(a) Signing of outgoing messages with own private key stored in a local storage.

(b) Verifying incoming messages using the attached sender's public key as well as verifying the attached sender's certificate itself.

(c) Encrypting outgoing messages with the receiver's public key.

(d) Decrypting incoming encrypted messages with own private key.

(e) Generating new private/public key pairs to be certified by a CA.

Since the module is intended to be compliant to IEEE 1609.2 [IEEE06], included components and units are according to respective standardization. The respective data flows and ECC operations are detailed in the following as well as, supplementary, summarized and visualized in Figure A.15.

Signing and Verifying

C2X messages are cryptographically signed based on current sender's pseudonym certificate. By subsequently verifying the certificate and the signature on receiver's

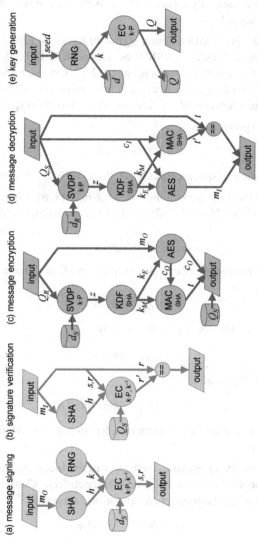

cf. [JSH11b]

FIGURE A.15.: Operations and data flow for ECC tasks, i.e., signing, verifying, encrypting, decrypting, and key generation.

side, *sender authenticity* and *message integrity* is ensured. Hence, stations which do not have certificated issued by a known CA as well as unauthorized manipulation of messages are detected.

C2X messages are signed with a digital signature based on *Elliptic Curve Digital Signature Algorithm* (ECDSA) [ANSI05] as detailed in [NIST09]. Thereby, the selected elliptic curves are defined over a prime field $\mathbb{F}_p$. Cryptography based on short term certificates is based on 244-bit curves in order to reduce number of transmitted bits. In contrast, long term certificate are issued for 256-bit cryptography.

The relevant curve parameters

$$(p, n, G, seed) \tag{A.2}$$

are selected according to specification NIST P-244 or NIST P-256, respectively.[73]

To generate a signature for an outgoing message m_O, a hash h has to be generated, advocating either SHA-224 or SHA-256 [NIST12] according to the selected elliptic curve.

$$h = \text{SHA}(m_O) \tag{A.3}$$

and a per message random number k is multiplied in EC arithmetic with the curve's base point G

$$(p_x, p_y) = k \cdot G \tag{A.4}$$

in a first step. Hence, the result r is achieved by

$$r \equiv p_x \quad \mod n \tag{A.5}$$

and, subsequently, the result s by

$$s \equiv k^{-1} \cdot (h + r \cdot d_S) \quad \mod n \quad , \tag{A.6}$$

whereas d_S is the private key of the current senders pseudonym. Hence, the signature is given by r and s.

Consequently, to verify a signature, the hash h of the incoming message's content m_I has to be generated on receiver's side, respectively. Combining h with the signature values r and s the numbers u_1 and u_2 are achieved.

$$u_1 \equiv h \cdot s^{-1} \quad \mod n \tag{A.7}$$

$$u_2 \equiv r \cdot s^{-1} \quad \mod n \tag{A.8}$$

---

[73] In fact, curve parameters of an elliptic curve according to NIST consist of additional parameters. However, to describe the ECC tasks in context of this thesis, only these parameters have to be regarded.

Subsequently, the EC operation

$$(p_x, p_y) = u_1 \cdot G + u_2 \cdot Q_S \quad , \tag{A.9}$$

whereas Q_S is the senders public key, leads to the preliminary result r' with

$$r' \equiv p_x \mod n \quad . \tag{A.10}$$

Thus, the signature is valid only if

$$r = r' \tag{A.11}$$

holds. Otherwise, either the message is not signed correctly or its content is changed after signing. However, in both cases the message has to be rejected.

A certificate's signature is verified analogous, with certificates content as m_I and issuing CA's public key as Q_S.

Encryption and Decryption

Generally, due to its broadcast nature, the content of a C2X messages is available to all road participants. However, for privacy reasons, the confidentiality of transmitted content may be ensured by encryption for particular message types.

IEEE 1609.2 specifies that C2X message content, if it is encrypted, it should be encrypted based on *Elliptic Curve Integrated Encryption Scheme* (ECIES) [IEEE04]. Thereby, the elliptic curve and its relevant parameters

$$(p, n, G, seed) \tag{A.12}$$

are given according to NIST P-256 over a prime field $\mathbb{F}_p$.

To encrypt an outgoing message's content m_O, a *Secret Value Derivation Primitive* (SVDP) based on EC multiplication according to [IEEE00][74] is advocated in a first step to achieved a shared secret z. Thereby, by taking a per message random number k and applying

$$(p_x, p_y) = d_S \cdot Q_R \quad , \tag{A.13}$$

whereas d_S is the senders private key and Q_R is the receivers public key.[75] Hence, the shared secret z is equivalent to the x-coordinate p_x of the EC multiplication result.

$$z \equiv p_x \mod n \quad . \tag{A.14}$$

---

[74] From the feasible choices given, *ECSVDP-DHC* is selected.

[75] The definition of the SVDP based on EC additionally involves the curve's cofactor. However, for elliptic curves according to NIST recommendation the cofactor is always one. Consequently it can be omitted.

Subsequently, a *Key Derivation Function* (KDF) as specified in [IEEE04][76] is advocated. Thereby, two keys k_E and k_M are achieved as the concatenated result generated by hashing the concatenation of z and the number 1 with SHA-256.

$$k_E \| k_M = \text{SHA}(z\|1) \tag{A.15}$$

The outgoing message's content m_O is encrypted with the 128 bit version of the symmetric cypher algorithm *Advanced Encryption Standard* (AES) [NIST01] in CCM mode [NIST04]. Hence, the cypher text c_O is calculated using the achieved symmetric key k_E.

$$c_O = \text{AES}_{k_E}(m_O) \tag{A.16}$$

A corresponding *Message Authentication Code* (MAC) t is generated regarding the achieved key k_M. Thereby, the *Hashed Message Authentication Code* (HMAC) based upon SHA-256 [NIST12] according to [NIST08] is advocated.

$$t = \text{MAC}_{k_M}(c_O) \tag{A.17}$$

Hence, the encrypted content c_O can be sent in combination with the MAC t and the sender's public key Q_S in a C2X message.

A received message with encrypted content c_I is decrypted analogous. In a first step, the shared secret z is achieved by providing the sender's public key Q_S and the receiver's private key d_R to the SVDP

$$(p_x, p_y) = d_R \cdot Q_S \tag{A.18}$$

$$z \equiv p_x \mod n \tag{A.19}$$

and, subsequently, advocating the KDF to achieve the secret keys k_E and k_M as described.

$$k_E \| k_M = \text{SHA}(z\|1) \tag{A.20}$$

Additionally, the MAC t' of the cypher text c_I is calculated on receivers side

$$t' = \text{MAC}_{k_M}(c_I) \quad . \tag{A.21}$$

Hence, message confidentiality and message integrity is only ensured if

$$t' = t \tag{A.22}$$

holds. Otherwise, the message has to be refused. Finally, the cypher text c_I is decrypted with the symmetric cypher algorithm AES using the symmetric key k_E

$$m_I = \text{AES}_{k_E}(c_I) \quad , \tag{A.23}$$

---

[76] From the feasible choices given, *KDF2* is selected.

such that message's plain text m_I is achieved.

Key Generation

For both, signing/verifying and encryption/decryption, elliptic curve private/public key pairs (d, Q) are mandatory. These keys are generated according to [NIST09] within the *Hardware Security Module* based on EC multiplication. The relevant curve parameters $(p, n, G, seed)$ are selected according to specification NIST P-244 or NIST P-256, respectively.

A random number k is generated, in a first step to achieve the EC key pair. This random number provides the private key d by

$$d \equiv (k \quad \mod (n-1)) + 1 \quad . \tag{A.24}$$

Accordingly, the public key Q, which is a point of the elliptic curve, is calculated with respect to the private key

$$Q = d \cdot G \quad , \tag{A.25}$$

such that a feasible EC private/public key pairs (d, Q) is achieved.

While the private key d should be stored in the local storage, the public key Q has to be sent to a corresponding CA for issuing a signed certificate.

A.3.2. The Module as a Pluggable USB Stick

The Hardware Security module shall be implemented on an FPGA board. This way, high adaptability of the core components is achieved. Additionally, for high compatibility the board is attachable via USB to a target C2X host system. Hence, it can easily be integrated into existing C2X communication systems.

Accordingly, within the basic module architecture, instructions and messages are delivered from the C2X host system serially via USB and processed by the *Communication Interface*. The control flow inside the Hardware Security module is handled by a *Controller* component. This controller distributes received data and keys from the Communication Interface via a local bus. Additionally, it activates according to the current queued instruction dedicated components for signing, verifying, encrypting, decrypting, or key generation.

After activation, the respective component takes over the control flow within the module. It consults special units for EC point multiplication (k·P), inversion (k^{-1}), hashing (SHA), symmetric encryption (AES), and a random number generator (RNG) on demand. This basic module architecture in terms of individual components and units as well as communication links between them is illustrated in Figure A.16.

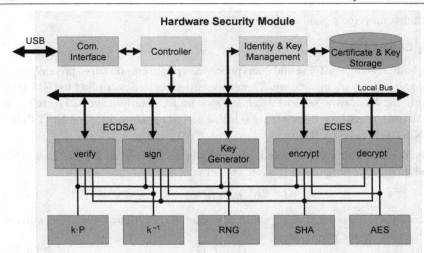

cf. [JSH11b]

FIGURE A.16.: Structure of the *Hardware Security* module, with respect to the implemented components and units as well as their interconnections.

Own public and private keys as well as buffered received public keys are accessible by request from the local storage. For message signing and decryption, the respective private key p_A is loaded from local certificate storage. In case of sending an encrypted message or verifying a received message, the respective external public key P_B (and certificate) is loaded into the module and is temporally stored in the local certificate storage. Thus, it has not to be loaded into the module again.

As mentioned, own short term certificates expire frequently and have to be replaced by new pseudonyms. Hence, a vehicle holds a pool of valid short term certificates for use on demand.

In order to achieve high confidentiality for private keys, the module provides facility to generate new keys. This way, in case that own certificates pool in local storage drains out, the module may generate new keys locally with use of the RNG. Subsequently, generated public keys are sent to a CA with a certification request. In contrast, respective generated private keys never leave the module. To identify the ITS station towards the CA for the certification request, a long term certificate is used. All these certificate management related tasks are handled by the component for *Identity & Key Management*.

The module is intended to provide a system as close to a final market system as possible. To achieve high compatibility and easy integration into a C2X communication framework, the module exploits the OpenSSL [OSSL] engine interface to provide its functionality. Hence, the module may relive CPU resources on C2X system exploiting cryptographic security build upon OpenSSL and ECC. However, to provide mentioned compatibility and adaptability to different research projects or future changes of standards, the design of the module does not aim towards highest performance, but, to meet hard space requirements on small FPGAs with an acceptable time-space trade-off.

Under supervision of the author of this thesis, in [ARZ11] a first prototypical prove of concept VHDL implementation of the presented Hardware Security module is provided. This implementation partly relies on sub-components developed in [EM11] and is designed for a *Virtex 5 ML505 Evaluation Platform*[77] [XILIN]. The design exploits about 69 % of the FPGAs resources and might be executed with a clock frequency of 40 MHz.

However, in [MAH15] a VHDL realization fitting on a smaller FPGA, e. g., *Xilinx Spartan-3* or *Altera Cyclon III* series, is realized under the supervision of the author of this thesis. Such low-cost FPGAs might be placed on evaluation boards with a size comparable with common USB flash drives and might be powered by USB, only. Since encryption and decryption is rarely needed, for this second implementation a focus is set on the most important parts, i. e., message signing and verification. Thus, encryption and decryption parts are omitted.

The implementation includes well-developed and highly optimized sub-components [MOL07] [RIE08] [LMR$^+$08] [LH08] [MAD02]. In addition, the design is made as flexible as possible such that it may be compiled for a various set of different FPGAs from different vendors. Moreover, in order to facilitate a comparison of multiple approaches, it might be switched from algorithms based on elliptic curves over $\mathbb{F}_p$ as stated in Section A.3.1 to an implementation realizing elliptic curve cryptography for comparable curves over $\mathbb{F}_{2^m}$.

Finally, the design is exemplary realized for the *BeMicro FPGA Evaluation Kit*[78] [ARROW]. Thereby, the final field arithmetic is realized with up to 6 pipeline stages, such that about 80 % of the FPGAs resources are allocated for the $\mathbb{F}_p$ version. This implementation might be executed with a clock frequency up to 71 MHz.

In contrast, the comparable version based on $\mathbb{F}_{2^m}$ ECC, exploits up to 9 pipeline stages within the final field arithmetic. Hence, the design utilizes almost the whole FPGA, with about 94 % occupation of the resources. Thereby, it still might be executed with a clock frequency of up to 77 MHz.

---

[77] The Virtex 5 ML505 Evaluation board hosts a Xilinx XC5VLX110T FPGA.
[78] The BeMicro FPGA Evaluation Kit hosts an Altera EP3C16F256C8N FPGA.

Operation	Throughput	
Generate Key Pair	4.1 – 4.5	keys/s
Encrypt Message	≈ 6	messages/s
Decrypt Message	≈ 6	messages/s
Generate Signature	3.75	signatures/s
Verify Signature	2	signatures/s

cf. [ARZ11]

TABLE A.1.: Summary of the performance provided by the prototypical implementation of Security Hardware module, realized on the Virtex 5 ML505 Evaluation Platform.

Operation	Throughput	
Generate Signature	113	signatures/s
Verify Signature	35	signatures/s

cf. [MAH15]

TABLE A.2.: Summary of the performance on $\mathbb{F}_p$-based ECC operations provided by an implementation of Security Hardware module, realized on the BeMicro FPGA Evaluation Kit.

A.3.3. Module Performance

Under the supervision of the author of this thesis, in [ARZ11] a prove of concept implementation of the Hardware Security module on a Virtex 5 board is developed. However, due to the prototypical realization, computational performance is very low and might not be able to sustainable support C2X systems as Table A.1 indicates.

Accordingly, a more advanced implementation is developed in [MAH15] under the supervision of the author of this thesis. As detailed in Section A.3.2, this version does not provide message encryption and decryption. However, an exemplary realization for a *BeMicro FPGA Evaluation Kit*, a small USB-based FPGA board with a size comparable to a USB flash drive, provides higher computational performance for $\mathbb{F}_p$ based ECC operations, as detailed in Table A.2

However, the design might be easily switched to ECC operations based on curves over $\mathbb{F}_{2^m}$. Hence, comparable tasks are performed, which leads to a significant higher throughput, as indicated in Table A.3.

However, even if these prototypical implementations do not reach the performance an ASIC implementation will reach, the tendency is clear. A Hardware Security module will significantly improve the performance of the C2X cryptography system and relieve substantial resources of the main processing unit.

Operation	Throughput	
Generate Signature	418	signatures/s
Verify Signature	216	signatures/s

cf. [MAH15]

TABLE A.3.: Summary of the performance on $\mathbb{F}_{2^m}$-based ECC operations provided by an implementation of Security Hardware module, realized on the BeMicro FPGA Evaluation Kit.

Moreover, the evaluation shows that, especially for the verification, ECC operations over $\mathbb{F}_{2^m}$ are by almost a magnitude faster than comparable operations over $\mathbb{F}_p$. Hence, the author of this thesis proposes to redesign or, at least, to extend respective standardization with $\mathbb{F}_{2^m}$-based cryptography. This way, overall security within C2X communication as well as gained safety related improvements might benefit from faster calculations.

Bibliography

[ACM⁺13] A. AUTOLITANO, C. CAMPOLO, A. MOLINARO, R. M. SCOPIGNO, and A. VESCO: *An insight into Decentralized Congestion Control techniques for VANETs from ETSI TS 102 687 V1.1.1*, 6th IEEE IFIP Wireless Days (WD'13), Valencia, Spain, November 2013

[ADAC11] ALLGEMEINER DEUTSCHER AUTOMOBIL-CLUB E. V.: *Mobilität in Deutschland – Ausgewählte Ergebnisse*, (German), February 2011

[ADAC12] ALLGEMEINER DEUTSCHER AUTOMOBIL-CLUB E. V.: *Aquaplaning – Unterschätzte Gefahr*, September 2012, http://www.adac.de/infotestrat/reifen/rund_um_den_reifen/aquaplaning/, (German)

[AKTIV] ADAPTIVE UND KOOPERATIVE TECHNOLOGIEN FÜR DEN INTELLIGENTEN VERKEHR (AKTIV): *Project Website*, http://www.aktiv-online.org

[AMU] AKTIVE MOBILE UNFALLVERMEIDUNG UND UNFALLFOLGENMINDERUNG DURCH KOOPERATIVE ERFASSUNGS- UND TRACKINGTECHNOLOGIE (AMULETT): *Project Website*, http://www.projekt-amulett.de/, (German)

[ANSI05] AMERICAN NATIONAL STANDARDS INSTITUTE: *Public Key Cryptography for the Financial Services Industry – The Elliptic Curve Digital Signature Algorithm (ECDSA)*, ANSI X9.62:2005, November 2005

[AOSA] AUTOMOTIVE OPEN SYSTEM ARCHITECTURE DEVELOPMENT COOPERATION: *Partnership Website*, http:\www.autosar.org

[ARROW] ARROW CENTRAL EUROPE GMBH: *BeMicro - Embedded Control Made Easy*, http://www.arroweurope.com/markets-solutions/solutions/bemicro/bemicro.html

[ARZ11] S. ARZT: *A Security Hardware Module for Elliptic Curve Cryptography in Car-to-Car Scenarios*, Master Thesis, TU Darmstadt, Supervisor: A. JAEGER, October 2011

[BAL05] C. A. BALANIS: *Antenna Theory: Analysis and Design*, John Wiley & Sons, May 2005

[BAST07]   D. KALINOWSKA, J. KLOAS, and H. KUHFELD: *Fahrerlaubnisbesitz in Deutschland*, Berichte der Bundesanstalt für Straßenwesen Heft M 187, (German), August 2007

[BB07]   A. BARISANI and A. BARISANI: *Injecting RDS-TMC Traffic Information Signals*, 8th CanSecWest – Applied Security Conference, Vancouver, British Columbia, Canada, April 2007

[BDL50]   LE BUREAU DES LONGITUDES: *Annuaire pour l'An 1950*, Gauthier-Villars, January 1950

[BER11]   Y. BERGHÖFER: *Implementation and Evaluation of a Car-2-X Framework on Facility Layer*, Bachelor Thesis, TU Darmstadt, Supervisors: H. STÜBING and B. BÜCHS, May 2011

[BH07]   L. BUTTYÁN and J.-P. HUBAUX: *Security and Cooperation in Wireless Networks: Thwarting Malicious and Selfish Behavior in the Age of Ubiquitous Computing*, Cambridge University Press, October 2007

[BK08]   R. BÜHLER and U. KUNERT: *Trends und Determinanten des Verkehrsverhaltens in den USA und in Deutschland*, (German), December 2008

[BNM12]   D. BIAL, S. NAGURNAS, and V. MITUNEVIČIUS: *Experimental research of car acceleration characteristics*, Technical Transactions – Mechanics, 9: 211–220, October 2012

[BON99]   J. A. BONNESON: *Side Friction and Speed as Controls for Horizontal Curve Design*, Journal of Transportation Engineering, 125: 473–480, November 1999

[BP99]   S. S. BLACKMAN and R. POPOLI: *Design and Analysis of Modern Tracking Systems*, Artech House, July 1999

[BS04]   A. R. BERESFORD and F. STAJANO: *Mix Zones: User Privacy in Location-aware Services*, 1st Annual IEEE Conference on Pervasive Computing and Communications (PerCom), Orlando, Florida, USA, March 2004

[BSM+09]   N. BISSMEYER, H. STÜBING, M. MATTHESS, J. P. STOTZ, J. SCHÜTTE, M. GERLACH, and F. FRIEDERICI: *sim^{TD} Security Architecture: Deployment of a Security and Privacy Architecture in Field Operational Tests*, 7th Conference Embedded Security in Cars (ESCAR 2009), Düsseldorf, Germany, November 2009

[C2C07]   CAR 2 CAR COMMUNICATION CONSORTIUM: *C2C-CC Manifesto*, August 2007

[C2C14]   CAR 2 CAR COMMUNICATION CONSORTIUM: *Trigger Conditions and Data Quality – Triggering Conditions: Adverse Weather Conditions*, January 2014

[CCD08]   R. CHIESA, S. CIAPPI, and S. DUCCI: *Profiling Hackers: The Science of Criminal Profiling as Applied to the World of Hacking*, Auerbach Publications, November 2008

[CCMM09]   Y. CAO, S. CASTRONOVO, A. MAHR, and C. MÜLLER: *On Timing and Modality Choice with Local Danger Warnings for Drivers*, 1st International Conference on Automotive User Interfaces and Interactive Vehicular Applications (AutomotiveUI 2009), Essen, Germany, September 2009

[CKJ12]   J. CASSELGREN, M. KUTILA, and M. JOKELA: *Slippery Road Detection by Using Different Methods of Polarised Light*, Advanced Microsystems for Automotive Applications 2012 – Smart Systems for Safe, Sustainable and Networked Vehicles, pages 207–220, June 2012

[CMC+10]   Y. CAO, A. MAHR, S. CASTRONOVO, M. THEUNE, C. STAHL, and C. MÜLLER: *Local Danger Warnings for Drivers: The Effect of Modality and level of Assistance on Driver Reaction.pdf*, 14th International ACM Conference on Intelligent User Interfaces (IUI), Hong Kong, China, February 2010

[CRE12]   J. CASSELGREN, S. ROSENDAHL, and J. ELIASSON: *Road Surface Information System*, 16th Standing International Road Weather Commission Conference (SIRWEC 2012), Helsinki, Finland, May 2012

[CTM10]   Y. CAO, M. THEUNE, and C. MÜLLER: *Multimodal Presentation of Local Danger Warnings for Drivers: A Situation-Dependent Assessment of Usability*, IEEE International Professional Communication Conference 2010 (IPCC), Enschede, The Netherlands, July 2010

[DDS10]   M. DAHL, S. DELAUNE, and G. STEEL: *Formal Analysis of Privacy for Vehicular Mix-Zones*, 15th European Symposium on Research in Computer Security (ESORICS 2010), Athens, Greece, September 2010

[DFKI]   DEUTSCHES FORSCHUNGSZENTRUM FÜR KÜNSTLICHE INTELLIGENZ GMBH: *Institute Website*, http://www.dfki.de/?set_language=en

[DH76]   W. DIFFIE and M. E. HELLMAN: *New Directions in Cryptography*, IEEE Transactions on Information Theory, 22:644–654, November 1976

[DMA89]   DEFENSE MAPPING AGENCY: *The Universal Grids: Universal Transverse Mercator (UTM) and Universeal Polar Stereographic (UPS)*, TM 8358.2, September 1989

[DRIVE]   DRIVE C2X CONSORTIUM: *Project Website*, http://www.drive-c2x.eu/

[DRIVE11a] DRIVE C2X CONSORTIUM: *Basic System Specification*, Internal Report IR23.1, June 2011

[DRIVE11b] DRIVE C2X CONSORTIUM: *Consolidated Description of Test Sites*, Internal Report IR33.3, September 2011

[DRIVE11c] DRIVE C2X CONSORTIUM: *Enhanced System Specification*, Deliverable D23.1, October 2011

[DRIVE12a] DRIVE C2X CONSORTIUM: DRIVE C2X *Vehicle Integration and Interoperability Check Report*, Deliverable D26.1, August 2012

[DRIVE12b] DRIVE C2X CONSORTIUM: *Enhanced System Software Manual*, Internal Report IR24.3, August 2012

[DRIVE12c] DRIVE C2X CONSORTIUM: *Revised System Specification*, Deliverable D23.1 v3.0, July 2012

[DRIVE13a] DRIVE C2X CONSORTIUM: DRIVE C2X *FOT Piloting*, Deliverable D34.1, June 2013

[DRIVE13b] DRIVE C2X CONSORTIUM: *Technical Evaluation*, Internal Report IR44.1, October 2013

[DRIVE14] DRIVE C2X CONSORTIUM: *Technical Performance of* DRIVE C2X *Functions in Full-Scale FOT Operations*, Deliverable D44.1, February 2014

[DWD] DEUTSCHER WETTER DIENST: *Institute Website*, http://www.dwd.de/, (German)

[EBS04] T. ECHAVEGUREN, M. BUSTOS, and H. de SOLMINIHAC: *A Method to Evaluate Side Friction in Horizontal Curves, Using Supply-Demand Concepts*, 6th International Conference on Managing Pavements: The Lessons, The Challenges, The Way Ahead, Brisbane, Queensland, Australia, October 2004

[EC10] EUROPEAN COMMISSION: *Commission regulation concerning type-approval requirements for windscreen wiper and washer systems of certain motor vehicles and implementing Regulation (EC) No 661/2009 of the European Parliament and of the Council concerning type-approval requirements for the general safety of motor vehicles, their trailers and systems, components and separate technical units intended*, Commission Regulation (EU) No 1008/2010, November 2010

[ECK13] C. ECKERT: *IT-Sicherheit: Konzepte – Verfahren – Protokolle*, Oldenbourg Wissenschaftsverlag, March 2013, (German)

[ECRSP] EUROPEAN COMMISSION: *EU Road Safety Policy*, http://ec.europa. eu/transport/road_safety/specialist/policy/

[ECSAD] EUROPEAN COMMISSION: *Statistics – Accidents Data*, http://ec. europa.eu/transport/road_safety/specialist/statistics/

[EK06] O. EGGERS and J. KRÜGER: *Verfahren und Vorrichtung zum Austauschen fahrzeugoriginärer Informationen*, DE102004039633 A1, (German), February 2006

[ELL83] D. ELLINGHAUS: *Wetter und Autofahren – Eine Untersuchung über den Einfluss des Wetters auf das Unfallgeschehen und die Verkehrssicherheit*, (German), November 1983

[EM11]   M. EMAMGHOLI and D. MÜNCH : *Sichere Hardware für Car-2-X Kummu-nikation*, Bachelor Thesis, TU Darmstadt, Supervisors: M. STÖTTINGER, (German), March 2011

[ESA06]   EUROPEAN SPACE AGENCY : *EGNOS – The European Geostationary Navigation Overlay System - A cornerstone of Galileo*, ESA SP-1303, December 2006

[ETSI09]   EUROPEAN TELECOMMUNICATIONS STANDARDS INSTITUTE : *Intelligent Transport Systems (ITS); Vehicular Communications; Basic Set of Applications; Definitions*, ETSI TR 102 638, June 2009

[ETSI10a]   EUROPEAN TELECOMMUNICATIONS STANDARDS INSTITUTE : *Intelligent Transport Systems (ITS); Communications Architecture*, ETSI EN 302 665, September 2010

[ETSI10b]   EUROPEAN TELECOMMUNICATIONS STANDARDS INSTITUTE : *Intelligent Transport Systems (ITS); Vehicular Communications; Basic Set of Applications; Part 1: Functional Requirements*, ETSI TS 102 637-1, September 2010

[ETSI10c]   EUROPEAN TELECOMMUNICATIONS STANDARDS INSTITUTE : *Intelligent Transport Systems (ITS); Vehicular Communications; Geographical Area Definition*, ETSI EN 302 931, December 2010

[ETSI10d]   EUROPEAN TELECOMMUNICATIONS STANDARDS INSTITUTE : *Intelligent Transport Systems (ITS); Vehicular communications; GeoNetworking; Part 2: Scenarios*, ETSI TS 102 636-2, March 2010

[ETSI11a]   EUROPEAN TELECOMMUNICATIONS STANDARDS INSTITUTE : *Intelligent Transport Systems (ITS); Decentralized Congestion Control Mechanisms for Intelligent Transport Systems operating in the 5 GHz range; Access layer part*, ETSI TS 102 687, July 2011

[ETSI11b]   EUROPEAN TELECOMMUNICATIONS STANDARDS INSTITUTE : *Intelligent Transport Systems (ITS); Vehicular Communications; Basic Set of Applications; Part 2: Specification of Cooperative Awareness Basic Service*, ETSI TS 102 637-2, March 2011

[ETSI11c]   EUROPEAN TELECOMMUNICATIONS STANDARDS INSTITUTE : *Intelligent Transport Systems (ITS); Vehicular communications; GeoNetworking; Part 4: Geographical addressing and forwarding for point-to-point and point-to-multipoint communications; Sub-part 1: Media-Independent Functionality*, ETSI TS 102 636-4-1, June 2011

[ETSI12]   EUROPEAN TELECOMMUNICATIONS STANDARDS INSTITUTE : *Intelligent Transport Systems (ITS); European profile standard for the physical and medium access control layer of Intelligent Transport Systems operating in the 5 GHz frequency band*, ETSI EN 302 663, November 2012

[ETSI13a]   EUROPEAN TELECOMMUNICATIONS STANDARDS INSTITUTE: *Intelligent Transport Systems (ITS); Users and applications requirements; Part 1: Facility layer structure, functional requirements and specifications*, ETSI TS 102 894-1, August 2013

[ETSI13b]   EUROPEAN TELECOMMUNICATIONS STANDARDS INSTITUTE: *Intelligent Transport Systems (ITS); Users and applications requirements; Part 2: Applications and facilities layer common data dictionary*, ETSI TS 102 894-2, June 2013

[ETSI13c]   EUROPEAN TELECOMMUNICATIONS STANDARDS INSTITUTE: *Intelligent Transport Systems (ITS); Vehicular Communications; Basic Set of Applications; Part 3: Specification of Decentralized Environmental Notification Basic Service*, ETSI EN 302 637-3, August 2013

[ETSI14]   EUROPEAN TELECOMMUNICATIONS STANDARDS INSTITUTE: *Intelligent Transport Systems (ITS); Facilities Layer Function, Part 2: Services Announcement*, ETSI DTS 102 890-2, (to be published in 2014)

[EVITA]   E-SAFETY VEHICLE INTRUSION PROTECTED APPLICATIONS: *Project Website*, http://www.evita-project.org/

[FAZ10]   F. RIEGER: *Vorratsdatenspeicherung - Du kannst dich nicht mehr verstecken*, Frankfurter Allgemeine Zeitung (F.A.Z.), (22.02.2010), February 2010, (German)

[FGSV08]   FORSCHUNGSGESELLSCHAFT FÜR STRASSEN- UND VERKEHRSWESEN E.V.: *Richtlinien für die Anlage von Autobahnen (RAA) – Ausgabe 2008*, FGSV-Verlag, May 2008

[FHWA09]   FEDERAL HIGHWAY ADMINISTRATION: *Speed Concepts: Informational Guide*, September 2009

[FOKUS]   FRAUNHOFER-INSTITUT FÜR OFFENE KOMMUNIKATIONSSYSTEME: *Institute Website*, https://www.fokus.fraunhofer.de/en/

[FRF+07]   J. FREUDIGER, M. RAYA, M. FÉLEGYHÁZI, P. PAPADIMITRATOS, and J.-P. HUBAUX: *Mix-Zones for Location Privacy in Vehicular Networks*, 1st ACM International Workshop on Wireless Networking for Intelligent Transportation Systems (WiN-ITS 2007), Vancouver, British Columbia, Canada, August 2007

[FSHS12]   J. FIRL, H. STÜBING, S. A. HUSS, and C. STILLER: *Predictive Maneuver Evaluation for Enhancement of Car-to-X Mobility Data*, 2012 IEEE Intelligent Vehicles Symposium, Alcalá de Henares, Spain, June 2012

[FSHS13]   J. FIRL, H. STÜBING, S. A. HUSS, and C. STILLER: *MARV-X: Applying Maneuver Assessment for Reliable Verification of Car-to-X Mobility Data*, IEEE Transactions on Intelligent Transportation Systems, 14: 1301–1312, September 2013

[GF09]   M. GERLACH and F. FRIEDERICI : *Implementing Trusted Vehicular Communications*, 69th IEEE Vehicular Technology Conference (VTC Spring 2009), Barcelona, Spain, April 2009

[GH05]   M. GRUTESER and B. HOH : *On the Anonymity of Periodic Location Samples*, International Conference on Security in Pervasive Computing (SPC '05), Boppard, Germany, April 2005

[HAR+09]   O. HENNIGER, L. APVRILLE, A. F. Y. ROUDIER, A. RUDDLE, and B. WEYL : *Security requirements for automotive on-board networks*, 9th IEEE International Conference on Telecommunications for Intelligent Transport Systems (ITST-2009), Lille, France, October 2009

[HAR12]   O. HARTKOPP : *The CAN networking subsystem of the Linux kernel*, 13th International CAN Conference (iCC), Neustadt an der Weinstraße, Germany, March 2012

[HHSV07]   A. HILLER, A. HINSBERGER, M. STRASSBERGER, and D. VERBURG : *Results from the WILLWARN Project*, 6th European Congress and Exhibition on Intelligent Transportation Systems and Services, Aalborg, Denmark, June 2007

[HJS11]   S. A. HUSS, A. JAEGER, and H. STÜBING : *sim^{TD}: Safe and Intelligent Mobility – Field Test Germany; Architecture and Applications for Car-to-Car Communication*, 48th ACM/EDAC/IEEE Design Automation Conference (DAC 2011): Workshop on Intra and Inter-Vehicle Networking: Past, Present, and Future, San Diego, California, USA, June 2011

[HLA05]   N. HAUTIÈRE, R. LABAYRADE, and D. AUBERT : *Estimation of the Visibility Distance by Stereovision: a Generic Approach*, 9th IAPR Conference on Machine Vision Applications, Tsukuba, Japan, May 2005

[HMS03]   C. HIDE, T. MOORE, and M. SMITH : *Adaptive Kalman Filtering for Low-cost INS/GPS*, Journal of Navigation, 56: 143–152, January 2003

[HTA07]   N. HAUTIÈRE, J.-P. TAREL, and D. AUBERT : *Simultaneous Contrast Restoration and Obstacles Detection: First Results*, 2007 IEEE Intelligent Vehicles Symposium, Istanbul, Turkey, June 2007

[IEC00]   INTERNATIONAL ELECTROTECHNICAL COMMISSION : *Specification of the radio data system (RDS) for VHF/FM sound broadcasting in the frequency range from 87.5 to 108.0 MHz*, IEC 62106:2000, January 2000

[IEEE00]   INSTITUTE OF ELECTRICAL AND ELECTRONICS ENGINEERS; COMPUTER SOCIETY : *Specifications for Public-Key Cryptography*, IEEE 1363™-2000, January 2000

[IEEE04]   INSTITUTE OF ELECTRICAL AND ELECTRONICS ENGINEERS; COMPUTER SOCIETY : *Specifications for Public-Key Cryptography – Amendment 1: Additional Techniques*, IEEE 1363a™-2004, September 2004

[IEEE06]   INSTITUTE OF ELECTRICAL AND ELECTRONICS ENGINEERS; INTELLI-
             GENT TRANSPORTATION SYSTEMS COMMITTEE : *Trial-Use Standard for
             Wireless Access in Vehicular Environments – Security Services for Applica-
             tions and Management Messages*, IEEE 1609.2™-2006, July 2006

[IEEE10]   INSTITUTE OF ELECTRICAL AND ELECTRONICS ENGINEERS : *Wireless
             LAN Medium Access Control (MAC) and Physical Layer (PHY) Specifications,
             Amendment 6: Wireless Access in Vehicular Environments*, IEEE 802.11p,
             July 2010

[ISO03a]   INTERNATIONAL ORGANIZATION FOR STANDARDIZATION : *Road vehicles
             – Controller Area Network (CAN) – Part 1: Data link layer and physical
             signalling*, ISO 11898-1, December 2003

[ISO03b]   INTERNATIONAL ORGANIZATION FOR STANDARDIZATION : *Road vehicles
             – Controller Area Network (CAN) – Part 2: High-speed medium access unit*,
             ISO 11898-2, December 2003

[ISO06]   INTERNATIONAL ORGANIZATION FOR STANDARDIZATION : *Road vehicles
             – Controller Area Network (CAN) – Part 3: Low-speed, fault-tolerant, medium-
             dependent interface*, ISO 11898-3, June 2006

[ISO09]   INTERNATIONAL ORGANIZATION FOR STANDARDIZATION : *Vehicle probe
             data for wide area communications*, ISO 22837:2009, January 2009

[ISO11]   INTERNATIONAL ORGANIZATION FOR STANDARDIZATION : *Road vehicles
             – Functional safety, Part 9: Automotive Safety Integrity Level (ASIL)-oriented
             and safety-oriented analyses*, ISO 26262-9:2011, November 2011

[ISO13]   INTERNATIONAL ORGANIZATION FOR STANDARDIZATION : *Intelligent
             transport systems – Traffic and travel information via transport protocol
             experts group, generation 1 (TPEG1) binary data format – Part 9: Traffic
             event compact (TPEG1-TEC)*, ISO TS 18234-9:2013, October 2013

[ITU09a]   INTERNATIONAL TELECOMMUNICATION UNION – TELECOMMUNICATION
             STANDARDIZATION SECTOR : *Information technology – Abstract Syntax No-
             tation One (ASN.1): Specification of basic notation*, ITU-T Recommendation
             X.680, December 2009

[ITU09b]   INTERNATIONAL TELECOMMUNICATION UNION – TELECOMMUNICATION
             STANDARDIZATION SECTOR : *Information technology – ASN.1 encoding
             rules: Specification of Basic Encoding Rules (BER), Canonical Encoding
             Rules (CER) and Distinguished Encoding Rules (DER)*, ITU-T Recommenda-
             tion X.690, December 2009

[IVSS10]   M. ANDERSSON, F. BRUZELIUS, J. CASSELGREN, M. HJORT, S. LÖFVING,
             G. OLSSON, J. RÖNNBERG, M. SJÖDAHL, S. SOLYOM, J. SVENDENIUS,
             and S. YNGVE : *Road Friction Estimation – Part II*, IVSS Project Report,
             November 2010

[JAMA]   MATHWORKS and NATIONAL INSTITUTE OF STANDARDS AND TECHNOL-
            OGY (NIST): *JAMA: A Java Matrix Package*, http://math.nist.gov/
            javanumerics/jama/

[JBSH12]   A. JAEGER, N. BISSMEYER, H. STÜBING, and S. A. HUSS: *A Novel Frame-
            work for Efficient Mobility Data Verification in Vehicular Ad-hoc Networks*,
            International Journal of Intelligent Transportation Systems Research (IJIR),
            10: 11–21, January 2012

[JF14]   A. JAEGER and H. FEIFEL: *Zugwarnvorrichtung*, Az.: 10 2014 19 321.4,
            Patent Pending, (German), September 2014

[JH11]   A. JAEGER and S. A. HUSS: *The Weather Hazard Warning in sim^{TD}: A
            Design for Road Weather Related Warnings in a Large Scale Car-to-X Field
            Operational Test*, 11th IEEE International Conference on Telecommunica-
            tions for Intelligent Transport Systems (ITST-2011), St. Petersburg, Russia,
            August 2011

[JH13]   A. JAEGER and S. A. HUSS: *Novel Techniques to Handle Rectangular Areas
            in Car-to-X Communication Applications*, 10th International Conference on
            Informatics in Control, Automation and Robotics (ICINCO 2013) – Special
            Session on Intelligent Vehicle Controls & Intelligent Transportation Systems
            (IVC&ITS 2013), Reykjavík, Iceland, July 2013

[JH15]   A. JAEGER and S. A. HUSS: *Verfahren zur Transformation einer Position-
            sangabe in ein lokales Koordinatensystem*, Az.: 10 2015 210 096.0, *Patent
            Pending*, (German), June 2015

[JÄN11]   M. JÄNSCH: *Entwurf und Implementierung einer Car-to-X-Komponente
            für Straßenwetterwarnungen*, Bachelor Thesis, TU Darmstadt, Supervisor:
            A. JAEGER, (German), June 2011

[JSH11a]   A. JAEGER, H. STÜBING, and S. A. HUSS: *A Dedicated Hardware Security
            Module for Field Operational Tests of Car-to-X Communication*, 4th ACM
            Conference on Wireless Network Security (WiSec'11), Hamburg, Germany,
            June 2011

[JSH11b]   A. JAEGER, H. STÜBING, and S. A. HUSS: *WiSec 2011 Poster: A Modular
            Design for a Hardware Security Module in Car-to-X Communication*, ACM
            SIGMOBILE Mobile Computing and Communications Review (MC2R),
            15: 43–44, November 2011

[KB12]   M. KLEINE-BUDDE: *SocketCAN – The official CAN API of the Linux ker-
            nel*, 13th International CAN Conference (iCC), Neustadt an der Weinstraße,
            Germany, March 2012

[KBFL^{+}11] F. KIRSCHKE-BILLER, S. FÜRST, S. LUPP, S. BUNZEL, R. RIMKUS, R.
            RIMKUS, A. GILBERG, K. NISHIKAWA, and A. TITZE: *AUTOSAR – A
            worldwide standard: Current developments, roll-out and outlook*, 15th In-

ternational VDI Congress "Electronic Systems for Vehicles", Baden-Baden, Germany, October 2011

[KCR⁺10]   K. KOSCHER, A. CZESKIS, F. ROESNER, S. PATEL, T. KOHNO, S. CHECK-OWAY, D. MCCOY, B. KANTOR, D. ANDERSON, H. SHACHAM, and S. SAV-AGE: *Experimental Security Analysis of a Modern Automobile*, 2010 IEEE Symposium on Security and Privacy, Oakland, California, USA, May 2010

[KGKM05]   H.-P. KRÜGER, M. GREIN, A. KAUSSNER, and C. MARK: *SILAB – A Task Oriented Driving Simulation*, Driving Simulation Conference North America 2005 (DSC-NA 2005), Orlando, Floriada, USA, December 2005

[KÁL60]   R. E. KÁLMÁN: *A New Approach to Linear Filtering and Prediction Problems*, Transactions of the American Society of Mechanical Engineers – Journal of Basic Engineering, 82: 35–45, March 1960

[KLA08]   F. KLANNER: *Entwicklung eines kommunikationsbasierten Querverkehrsassistenten im Fahrzeug*, VDI-Verlag, May 2008, (German)

[KRCS08]   C. KEIL, A. RÖPNACK, G. C. CRAIG, and U. SCHUMANN: *Sensitivity of quantitative precipitation forecast to height dependent changes in humidity*, Geophysical Research Letters, 35: 1–5, May 2008

[KSSH12]   B. KLOIBER, T. STRANG, H. SPIJKER, and G. HEIJENK: *Improving Information Dissemination in Sparse Vehicular Networks by Adding Satellite Communication*, IEEE Intelligent Vehicles Symposium (IV'2011), Alcalá de Henares, Spain, June 2012

[KW11]   H. KRISHNAN and A. WEIMERSKIRCH: *"Verify-on-Demand" – A Practical and Scalable Approach for Broadcast Authentication in Vehicle-to-Vehicle Communication*, SAE International Journal of Passenger Cars – Mechanical Systems, 4: 536–546, June 2011

[LB09]   C. LAURENDEAU and M. BARBEAU: *Probabilistic Localization and Tracking of Malicious Insiders Using Hyperbolic Position Bounding in Vehicular Networks*, EURASIP Journal on Wireless Communications and Networking – Special issue on wireless network security, 2009: 1-12, May 2009

[LBV07]   T. H. LEVENTE BUTTYÁN, , and I. VAJDA: *On the Effectiveness of Changing Pseudonyms to Provide Location Privacy in VANETs*, 4th European Workshop on Security and Privacy in Ad-hoc and Sensor Networks (ESAS '07), Cambridge, UK, July 2007

[LEKK09]   P. A. LEWANDOWSKI, W. E. EICHINGER, A. KRUGER, and W. F. KRAJEWSKI: *Lidar-Based Estimation of Small-Scale Rainfall: Empirical Evidence*, Journal of Atmospheric and Oceanic Technology, 26: 656–664, March 2009

[LH08]   R. LAUE and S. A. HUSS: *Parallel Memory Architecture for Elliptic Curve Cryptography over $\mathbb{GF}(p)$ Aimed at Efficient FPGA Implementation*, Journal of Signal Processing Systems for Signal, Image, and Video Technology, 51: 39–55, April 2008

[LHLS12] J. LAURENT, J. F. HÉBERT, D. LEFEBVRE, and Y. SAVARD: *Using 3D Laser Profiling Sensors for the Automated Measurement of Road Surface Conditions*, 7th RILEM International Conference on Cracking in Pavements, Delft, The Netherlands, June 2012

[LMR+08] R. LAUE, H. G. MOLTER, F. RIEDER, K. SAXENA, and S. A. HUSS: *A Novel Multiple Core Co-Processor Architecture for Efficient Server-based Public Key Cryptographic Applications*, IEEE Computer Society Annual Symposium on VLSI, Montpellier, France, April 2008

[LSK06] T. LEINMÜLLER, E. SCHOCH, and F. KARGL: *Position Verification Approaches for Vehicular Ad Hoc Networks*, IEEE Wireless Communications, 13: 16–21, October 2006

[LSS+08] T. LEINMÜLLER, R. K. SCHMIDT, E. SCHOCH, A. HELD, and G. SCHÄFER: *Modeling Roadside Attacker Behavior in VANETs*, IEEE Global Communications Conference (GLOBECOM) 2008 – 3rd IEEE Automotive Networking and Applications (AutoNet) Workshop, New Orleans, Louisiana, USA, December 2008

[LU07] J. J. LU: *Vehicle Traction Performance on Snowy and Icy Surfaces*, Transportation Research Record: Journal of the Transportation Research Board, 1536: 82–89, January 2007

[MAD02] F. MADLENER: *FPGA based Hardware Acceleration for Elliptic Curve Cryptography based on* $\mathbb{GF}(2^n)$, Student Thesis, TU Darmstadt, Supervisor: M. ERNST, August 2002

[MAH15] R. MAHS: *Hardwarebeschleunigung für Digitale Signaturen auf Basis von Elliptischen Kurven in der Car-to-Car-Kommunikation*, Diploma Thesis, TU Darmstadt, Supervisor: A. JAEGER, (German), October 2015

[MAI04] C. MAIHÖFER: *A Survey of Geocast Routing Protocols*, IEEE Communications Surveys & Tutorials, 6: 32–42, April 2004

[MEE98] J. MEEUS: *Astronomical Algorithms*, Willmann-Bell, December 1998

[METEO] METEOMEDIA AG: *Website*, http://www.mminternational.com

[MIG09] S. MIGLIETTA: *Entwicklung eines Simulators zur Evaluation des Adaptive Beamforming in der Car-to-X Kommunikation*, Bachelor Thesis, Hochschule Darmstadt, Supervisor: H. STÜBING, (German), August 2009

[MOL07] H. G. MOLTER: *Flexibler Krypto-CoProzessor für Server als SoC*, Diploma Thesis, TU Darmstadt, Supervisor: R. LAUE, (German), May 2007

[MSS10] M.-R. MOSAVI, M. SADEGHIAN, and S. SAEIDI: *Improvement in Differential GPS Accuracy Using Kalman Filter*, Journal of Aerospace Science & Technology, 7: 139–150, October 2010

[NHTSA11] NATIONAL HIGHWAY TRAFFIC SAFETY ADMINISTRATION : *Vehicle Safety Communications – Applications (VSC-A) – Final Report: Appendix Volume 1 System Design and Objective Test*, September 2011

[NIMA00] NATIONAL IMAGERY AND MAPPING AGENCY : *World Geodetic System 1984 – Its Definition and Relationships with Local Geodetic Systems*, NIMA TR 8350.2, January 2000

[NIST01] NATIONAL INSTITUTE OF STANDARDS AND TECHNOLOGY : *Advanced Encryption Standard (AES)*, FIPS PUB 197, November 2001

[NIST04] NATIONAL INSTITUTE OF STANDARDS AND TECHNOLOGY : *Recommendation for Block Cipher Modes of Operation: The CCM Mode for Authentication and Confidentiality*, NIST SP 800-38C, May 2004

[NIST08] NATIONAL INSTITUTE OF STANDARDS AND TECHNOLOGY : *The Keyed-Hash Message Authentication Code (HMAC)*, FIPS PUB 198-1, July 2008

[NIST09] NATIONAL INSTITUTE OF STANDARDS AND TECHNOLOGY : *Digital Signature Standard (DSS)*, FIPS PUB 186-3, June 2009

[NIST12] NATIONAL INSTITUTE OF STANDARDS AND TECHNOLOGY : *Secure Hash Standard (SHS)*, FIPS PUB 180-4, June 2012

[OBL04] A. F. ORLANDO, J. D. BRIONIZIO, and L. A. LIMA : *Calculation of humidity parameters and uncertainties using different formulations and softwares*, 9th Symposium on Temperature and Thermal Measurements in Industry and Science (TEMPMEKO), Dubrovnik-Cavtat, Croatia, June 2004

[OSSL] THE OPENSSL PROJECT : *Project Website*, http://www.openssl.org/

[PM07] K. R. PETTY and W. P. MAHONEY III : *Enhancing road weather information through Vehicle Infrastructure Integration (VII)*, Transportation Research Record: Journal of the Transportation Research Board, 2015: 132–140, January 2007

[PMC+08] K. R. PETTY, W. P. MAHONEY III, J. R. COWIE, A. P. DUMONT, and W. L. MYERS : *Providing Winter Road Maintenance Guidance: Update of the Federal Highway Administration Maintenance Decision Support System (MDSS)*, 7th International Symposium on Snow Removal and Ice Control Technology, Indianapolis, Indiana, USA, June 2008

[POR11] B. E. PORTER (Ed.) : *Handbook of Traffic Psychology*, Academic Press, September 2011

[PREDR09] PRE-DRIVE C2X CONSORTIUM : *Refined Architecture*, Deliverable D1.2, August 2009

[QUA11] D. QUANZ : *Implementierung einer Fahrzeugplausibilitätsprüfung basierend auf Kommunikations- und Sensordaten*, Bachelor Thesis, TU Darmstadt, Supervisors: N. BISSMEYER and A. JAEGER, (German), January 2011

[RH05]   M. RAYA and J.-P. HUBAUX : *The Security of Vehicular Ad Hoc Networks*, 3rd ACM Workshop on Security of Ad Hoc and Sensor Networks, Alexandria, Virginia, USA, November 2005

[RIE08]   F. RIEDER : *Implementierung von modularer Multiplikation mit kurzen Pipelines für einen flexiblen Krypto-CoProzessor*, Diploma Thesis, TU Darmstadt, Supervisor: R. LAUE, (German), January 2008

[RITA]   RESEARCH AND INNOVATIVE TECHNOLOGY ADMINISTRATION : *Connected Vehicle Research Website*, http://www.its.dot.gov/connected_vehicle/connected_vehicle.htm

[RITA10]   RESEARCH AND INNOVATIVE TECHNOLOGY ADMINISTRATION – INTELLIDRIVE^{SM} : *Enabling Aftermarket Devices with DSRC-Based Communications Capabilities: Summary of Input from Industry Stakeholders*, June 2010

[RLY+09]   Z. REN, W. LI, Q. YANG, S. WU, and L. CHEN : *Location Security in Geographic Ad hoc Routing for VANETs*, International Conference on Ultra Modern Telecommunications (ICUMT '09), St. Petersburg, Russia, October 2009

[RS07]   S. REZAEI and R. SENGUPTA : *Kalman filter based integration of DGPS and vehicle sensors for localization*, IEEE Transactions on Control Systems Technology, 15: 1080–1088, November 2007

[RUP11]   T. RUPPENTHAL : *Entwicklung eines Simulators zur Evaluation des Secure Beamforming in der Car-to-X Kommunikation*, Bachelor Thesis, TU Darmstadt, Supervisor: H. STÜBING, (German), May 2011

[SBH+10]   H. STÜBING, M. BECHLER, D. HEUSSNER, T. MAY, I. RADUSCH, H. RECHNER, and P. VOGEL : *sim^{TD}: A Car-to-X System Architecture For Field Operational Tests*, IEEE Communications Magazine – Automotive Networking Series, 48: 148–154, May 2010

[SCH10]   D. SCHMIDT : *Fehleranalyse und Datenfusion von Satellitennavigations- und Fahrdynamiksensorsignalen*, VDI-Verlag, April 2010, (German)

[SÇH11]   H. STÜBING, M. ÇEVEN, and S. A. HUSS : *A Diffie-Hellman based Privacy Protocol for Car-to-X Communication*, 9th IEEE Conference on Privacy, Security and Trust (PST 2011), Montreal, Quebec, Canada, July 2011

[SCN14]   SAFE CAR NEWS : *US: Cadillac to launch "Intelligent and Connected" safety tech from 2017*, September 2014

[SEI08]   K. L. SEITTER : *Letter to P. R. Brubaker, U.S. Department of Transportation*, July 2008

[simTD]   SICHERE INTELLIGENTE MOBILITÄT – TESTFELD DEUTSCHLAND : *Project Website*, http://www.simtd.de/index.dhtml/enEN/

[SIMTD09a]   SICHERE INTELLIGENTE MOBILITÄT – TESTFELD DEUTSCHLAND : *An-forderungen der Funktionen an die Gesamtarchitektur*, Deliverable D11.4, (German), October 2009

[SIMTD09b]   SICHERE INTELLIGENTE MOBILITÄT – TESTFELD DEUTSCHLAND : *Aus-gewählte Funktionen*, Deliverable D11.2, (German), June 2009

[SIMTD09c]   SICHERE INTELLIGENTE MOBILITÄT – TESTFELD DEUTSCHLAND : *Be-wertende Übersicht existierender Systemarchitekturen*, Deliverable D21.1, (German), June 2009

[SIMTD09d]   SICHERE INTELLIGENTE MOBILITÄT – TESTFELD DEUTSCHLAND : *For-male Spezifikation des HMI*, Deliverable D22.2, (German), June 2009

[SIMTD09e]   SICHERE INTELLIGENTE MOBILITÄT – TESTFELD DEUTSCHLAND : *Funk-tionsspezifikation*, Deliverable D11.3, (German), September 2009

[SIMTD09f]   SICHERE INTELLIGENTE MOBILITÄT – TESTFELD DEUTSCHLAND : *Gro-barchitektur*, Working Document W21.2, (German), May 2009

[SIMTD09g]   SICHERE INTELLIGENTE MOBILITÄT – TESTFELD DEUTSCHLAND : *Kon-solidierter Systemarchitekturentwurf*, Deliverable D21.2, (German), October 2009

[SIMTD09h]   SICHERE INTELLIGENTE MOBILITÄT – TESTFELD DEUTSCHLAND : *Of-fene, ausschreibungsfähige Spezifikation der IRS*, Deliverable D22.2, (German), September 2009

[SIMTD09i]   SICHERE INTELLIGENTE MOBILITÄT – TESTFELD DEUTSCHLAND : *Spezi-fikation der IT-Sicherheitslösung*, Deliverable D21.5, (German), October 2009

[SIMTD09j]   SICHERE INTELLIGENTE MOBILITÄT – TESTFELD DEUTSCHLAND : *Spez-ifikation der Kommunikationsprotokolle*, Deliverable D21.4, (German), September 2009

[SIMTD09k]   SICHERE INTELLIGENTE MOBILITÄT – TESTFELD DEUTSCHLAND : *Spezi-fikation der sim^{TD} Funkschnittstellen*, Deliverable D21.3, (German), September 2009

[SIMTD09l]   SICHERE INTELLIGENTE MOBILITÄT – TESTFELD DEUTSCHLAND : *Test-und Versuchsspezifikation*, Working Document W13.1, (German), October 2009

[SIMTD09m]   SICHERE INTELLIGENTE MOBILITÄT – TESTFELD DEUTSCHLAND : *Un-falldatenanalyse – GIDAS-Wirkfeldanalyse ausgewählter sim^{TD}-Anwendungs-fälle zur Darstellung eines maximal anzunehmenden Wirkfeldes*, Deliverable D5.3 – Teil 2, (German), September 2009

[SIMTD09n]   SICHERE INTELLIGENTE MOBILITÄT – TESTFELD DEUTSCHLAND : *Vali-dierungsziele*, Deliverable D12.1, (German), September 2009

[SIMTD10a]   SICHERE INTELLIGENTE MOBILITÄT – TESTFELD DEUTSCHLAND : *Fahr-zeugseitiges IT-Sicherheitssystem*, Deliverable D22.3, (German), March 2010

[SIMTD10b]   SICHERE INTELLIGENTE MOBILITÄT – TESTFELD DEUTSCHLAND : *Spezifikation der IVS*, Deliverable D22.1, (German), February 2010

[SIMTD10c]   SICHERE INTELLIGENTE MOBILITÄT – TESTFELD DEUTSCHLAND : *Test- und Versuchsspezifikation*, Deliverable D13.2, (German), April 2010

[SIMTD10d]   SICHERE INTELLIGENTE MOBILITÄT – TESTFELD DEUTSCHLAND : *Versuchsplan 1.0*, Deliverable D44.1, (German), June 2010

[SIMTD12a]   SICHERE INTELLIGENTE MOBILITÄT – TESTFELD DEUTSCHLAND : *Implementierung und Dokumentation der fahrzeugseitigen Funktionen*, Deliverable D22.4, (German), January 2012

[SIMTD12b]   SICHERE INTELLIGENTE MOBILITÄT – TESTFELD DEUTSCHLAND : *Messtechnik*, Deliverable D24.2a, (German), January 2012

[SIMTD13a]   SICHERE INTELLIGENTE MOBILITÄT – TESTFELD DEUTSCHLAND : *Beschreibung und Auswertung des Anwendungsfalls A_2.1.3.1 Warnung vor Wettergefahren in der Fahrsimulation*, addendum to Working Document W43.1, (German), July 2013

[SIMTD13b]   SICHERE INTELLIGENTE MOBILITÄT – TESTFELD DEUTSCHLAND : *Ergebnisse Fahrsimulation*, Working Document W43.1, (German), June 2013

[SIMTD13c]   SICHERE INTELLIGENTE MOBILITÄT – TESTFELD DEUTSCHLAND : *Ergebnisse Feldversuch*, Working Document W43.2, (German), February 2013

[SIMTD13d]   SICHERE INTELLIGENTE MOBILITÄT – TESTFELD DEUTSCHLAND : *Simulationsumgebung V2.0 – Fahrsimulation*, Working Document W41.3a, (German), January 2013

[SIMTD13e]   SICHERE INTELLIGENTE MOBILITÄT – TESTFELD DEUTSCHLAND : *Technische Auswertung: F_1.1.3 – Erfassung der Straßenwetterlage*, addendum to Working Document W43.2, (German), August 2013

[SIMTD13f]   SICHERE INTELLIGENTE MOBILITÄT – TESTFELD DEUTSCHLAND : *Technische Auswertung: F_2.1.3 – Straßenwetterwarnung*, addendum to Working Document W43.2, (German), August 2013

[SJ10]   H. STÜBING and A. JAEGER : *Secure Beamforming for Weather Hazard Warning Application in Car-to-X Communication*, Lecture Notes in Electrical Engineering – Design Methodologies for Secure Embedded Systems, 78: 187–206, November 2010

[SJB+10]   H. STÜBING, A. JAEGER, N. BISSMEYER, C. SCHMIDT, and S. A. HUSS : *Verifying Mobility Data under Privacy Considerations in Car-to-X Communication*, 17th ITS World Congress, Busan, South Korea, October 2010

[SJWH11a]   H. STÜBING, A. JAEGER, N. WAGNER, and S. A. HUSS : *Integrating Secure Beamforming into Car-to-X Architectures*, SAE International Journal of Passenger Cars – Electronic and Electrical Systems, 4: 88–96, June 2011

[SJWH11b]   H. STÜBING, A. JAEGER, N. WAGNER, and S. A. HUSS: *Integrating Secure Beamforming into Car-to-X Architectures*, SAE 2011 World Congress & Exhibition, Detroit, Michigan, USA, April 2011

[SKL+06]   E. SCHOCH, F. KARGL, T. LEINMÜLLER, S. SCHLOTT, and P. PAPADIMITRATOS: *Impact of pseudonym changes on geographic routing in VANETs*, 3rd European conference on Security and Privacy in Ad-Hoc and Sensor Networks (ESAS '06), Hamburg, Germany, September 2006

[SLS+08]   R. K. SCHMIDT, T. LEINMÜLLER, E. SCHOCH, A. HELD, and G. SCHÄFER: *Vehicle Behavior Analysis to Enhance Security in VANETs*, 4th Workshop on Vehicle to Vehicle Communications (V2VCOM), Eindhoven, the Netherlands, July 2008

[SMI02]   M. SMITH: *Vehicle-Centric Weather Prediction System and Method*, US Pat. 2002/0067289 A1, June 2002

[SMK09]   F. SCHAUB, Z. MA, and F. KARGL: *Privacy Requirements in Vehicular Communication Systems*, 3rd International Conference on Computational Science and Engineering (CSE 2009), Vancouver, British Columbia, Canada, August 2009

[SMM+05]   W. SPECKS, K. MATHEUS, R. MORICH, C. MENIG, and I. PAULUS: *CAR 2 CAR Communication – Market Introduction and Success Factors*, 5th European Congress and Exhibition on Intelligent Transport Systems and Services (European ITS 2005), Hannover, Germany, June 2005

[SNY87]   J. P. SNYDER: *Map Projections — A Working Manual*, US Geologic Survey Professional Paper, 1395, November 1987

[SOCAN]   SOCKETCAN: *Project Website*, https://gitorious.org/linux-can

[SON90]   D. SONNTAG: *Important new Values of the Physical Constants of 1986, Vapour Pressure Formulations based on ITS-90, and Psychrometer Formulae*, Zeitschrift für Meteorologie, 40: 340–344, October 1990

[SPH11]   H. STÜBING, M. PFALZGRAF, and S. A. HUSS: *A Decentralized Group Privacy Protocol for Vehicular Networks*, 3th IEEE International Conference on Information Privacy, Security, Risk and Trust (PASSAT 2011), Boston, Massachusetts, USA, October 2011

[SPTG09]   B. L. SMITH, B. B. PARK, H. TANIKELLA, and N. J. GOODALL: *Preparing to Use Vehicle Infrastructure Integration (VII) in Transportation Operations: Phase II*, Transportation Research Board, April 2009

[SSH09]   H. STÜBING, A. SHOUFAN, and S. A. HUSS: *A Secure C2X Communication System based on Adaptive Beamforming*, 14th International VDI Congress "Electronic Systems for Vehicles", Baden-Baden, Germany, October 2009

[SSH10a]   H. STÜBING, A. SHOUFAN, and S. A. HUSS: *A Demonstrator for Beamform-*

ing in C2X Communication, 3rd IEEE International Symposium on Wireless Vehicular Communications (WiVeC 2010), Taipei, Taiwan, May 2010

[SSH10b] H. STÜBING, A. SHOUFAN, and S. A. HUSS : *Enhancing Security and Privacy in C2X Communication by Radiation Pattern Control*, 3rd IEEE International Symposium on Wireless Vehicular Communications (WiVeC 2010), Taipei, Taiwan, May 2010

[SWOV12] INSTITUTE FOR ROAD SAFETY RESEARCH : *SWOV Fact sheet: The influence of weather on road safety*, February 2012

[SYM10] D. SYMEONIDIS : *RDS-TMC Spoofing using GNU Radio*, 6th Karlsruhe Workshop on Software Radios, Karlsruhe, Germany, March 2010

[TAR10] C. TARNOVSKY : *Deconstructing a 'Secure' Processor*, Black Hat DC 2010, Washington D. C., USA, February 2010

[TMSH06] M. TORRENT-MORENO, P. SANTI, and H. HARTENSTEIN : *Distributed Fair Transmit Power Adjustment for Vehicular Ad Hoc Networks*, 3rd Annual IEEE Communications Society Conference on Sensor, Mesh and Ad Hoc Communications and Networks (SECON '06), Reston, Virginia, USA, September 2006

[TPEG06] M. DREHER, G. OBERT, J. MERTZ, A. RUDOLF, M. SCHOLZ, R. van den BERG, M. HESSLING, H.-W. PFEIFFER, M. LÄDKE, and I. PETROV : *TPEG TEC Application Specification*, Mobile.Info, March 2006

[TUR97] M. TURUNEN : *Measuring salt and freezing temperature on roads*, Meteorological Applications, 4: 11–15, March 1997

[WEN07] J. WENDEL : *Integrierte NavigatNavigation – Sensordatenfusion, GPS und Inertiale Navigation*, Oldenbourg Wissenschaftsverlag, February 2007, (German)

[WIVW] WÜRZBURGER INSTITUT FÜR VERKEHRSWISSENSCHAFTEN GMBH : *Institute Website*, http://www.wivw.de/

[WMKP10] B. WIEDERSHEIM, Z. MA, F. KARGL, and P. PAPADIMITRATOS : *Privacy in Inter-Vehicular Networks: Why simple pseudonym change is not enough*, 7th International Conference on Wireless On-demand Network Systems and Services (WONS), Kranjska Gora, Slovenia, February 2010

[XILIN] XILINX, INC. : *Virtex-5 LXT FPGA ML505 Evaluation Platform*, http://www.xilinx.com/products/boards-and-kits/hw-v5-ml505-uni-g.html

[XYG06] B. XIAO, B. YU, and C. GAO : *Detection and Localization of Sybil Nodes in VANETs*, 12th Annual International Conference on Mobile Computing and Networking (MobiCom) – Workshop on Dependability Issues in Wireless Ad Hoc Networks and Sensor Networks (DIWANS '06), Los Angeles, California, USA, September 2006

[YCWO07] G. YAN, G. CHOUDHARY, M. C. WEIGLE, and S. OLARIU: *Providing VANET Security Through Active Position Detection*, 4th ACM International Workshop on Vehicular Ad Hoc Networks (VANET 2007), Montréal, Québec, Canada, September 2007

[YMM13] B. YING, D. MAKRAKIS, and H. T. MOUFTAH: *Dynamic Mix-Zone for Location Privacy in Vehicular Networks*, IEEE Communications Letters, 17: 1524–1527, July 2013

[ZOG13] J.-M. ZOGG: *GPS und GNSS – Grundlagen der Ortung und Navigation mit Satelliten*, (German), August 2013

[ZSGW09] Y. ZANG, S. SORIES, G. GEHLEN, and B. WALKE: *Towards a European Solution for Networked Cars – Integration of Car-to-Car technology into cellular systems for vehicular communication in Europe*, Geneva International Motor Show – The Fully Networked Car Workshop 2009, Geneva, Switzerland, March 2009

Printed in the United States
By Bookmasters